X-ray Crystallography

X-ray Crystallography

SECOND EDITION

William Clegg

UNIVERSITY PRESS

UNIVERSITY PRESS

Great Clarendon Street, Oxford OX2 6DP,
United Kingdom

Oxford University Press is a department of the University of Oxford.
It furthers the University's objective of excellence in research, scholarship,
and education by publishing worldwide. Oxford is a registered trade mark of
Oxford University Press in the UK and in certain other countries

Published in the United States of America by Oxford University Press
198 Madison Avenue, New York, NY 10016, United States of America

British Library Cataloguing in Publication Data

Data available

Library of Congress Control Number: 2014957583

ISBN 978-0-19-870097-5

Printed in Great Britain by Ashford Colour Press Ltd, Gosport, Hampshire

Preface to second edition

I am pleased that my 1998 Oxford Chemistry Primer, entitled *Crystal Structure Determination*, continues to be recommended and used as a standard text in many undergraduate chemistry courses around the world. After 16 years, however, it is really quite dated. One of the reasons why X-ray crystallography is an important topic to present to chemistry students alongside other structural techniques is that it plays such a major role in modern research, supplying otherwise unobtainable information about the details of both molecular and extended network structures in the solid state. As a wide-ranging research field in its own right, crystallography constantly sees enormous developments in the understanding of its fundamental principles and in their practical application across many physical and biological sciences as well as engineering. Appropriately in this International Year of Crystallography (IYCr 2014), I am grateful to Alice Roberts and others at OUP for the invitation to provide a revised, expanded, and updated edition, with the new title of *X-ray Crystallography*.

Of the original four chapters, the first has seen the least change, as the fundamental principles of X-ray crystallography remain as they were; I have mostly replaced the various examples, and have provided an updated account of X-ray sources. Chapter 2 has a shift in balance from older photographic and serial-diffractometer experiments to area detectors that are now routinely used and are being further developed; some newer methods of solving and refining crystal structures have been included, together with a brief account of some of the problems often encountered and how they are dealt with; and the reporting and archiving of crystal structures are described here. I have replaced all but one of the case studies in Chapter 3 by more recent examples, ensuring that all the main points of Chapter 2 are illustrated by at least one example. Here and elsewhere, full references are given to the example structures, all of which have been published; the data and results are being made available online so that they can be further investigated by students and teachers. The related topics of neutron diffraction and powder diffraction, retained largely unchanged in Chapter 4, have been supplemented by an account of biological macromolecular crystallography and a brief description of crystal structure prediction. As with other second-edition Primers, an extensive glossary of terms is provided. Each chapter has exercises for the reader, answers to which are available online along with further exercises.

I thank the OUP editorial staff for their advice, assistance, and patience, my colleagues and collaborators for the research projects that have provided examples and case studies in this book, and Dr Ehmke Pohl of Durham University for his suggestions for improvement of the section on macromolecular crystallography.

Bill Clegg
Newcastle upon Tyne
September 2014

Contents

Fundamentals of X-ray crystallography

1.1 Introduction

This book aims to provide a basic introduction to the technique of structure determination by X-ray **crystallography** at the chemistry undergraduate level. It is not intended as a detailed practical manual for researchers in the subject. The approach taken is to introduce fundamental principles and concepts, then to show in outline how these are used in practice, and to provide a number of case studies by way of illustration. A few related topics are discussed in the final chapter.

This first chapter describes the importance of X-ray crystallography in the context of modern chemistry, explains its basis by an optical analogy, outlines the main relevant properties of crystalline materials, especially aspects of symmetry, and provides a basic description of diffraction phenomena. Three important properties of the diffraction pattern of a **single crystal** are examined and related to features of interest in the crystal structure: the geometry of diffraction, symmetry observed in the pattern, and the variation of intensity in the discrete diffraction measurements; mathematical details are kept to a minimum, and are illustrated graphically as well as being explained in words. Finally, a brief description is given of the available sources of X-rays for crystallography.

Most of the structures used as illustrations have been published and are recorded in the Cambridge Structural Database (CSD; **databases** are discussed in Chapter 2); in each case a literature reference is given, together with the CSD **REFCODE**, and computer results and data files are available online, so that the structures can be investigated further by readers and teachers.

1.2 Crystallography compared with other structural techniques

A knowledge of the structure of both molecular and non-molecular materials is one of the fundamental aims of chemistry and is essential for a proper understanding of the physical and chemical properties of the materials. The term 'structure' has many meanings; here we take it to be the relative positions of the atoms or ions which make up the substance under study and hence a geometrical description in terms of bond lengths and angles, torsion angles and other measures of conformation, **intramolecular** and **intermolecular** non-bonded distances and interactions such as **hydrogen bonding**, and other quantities of interest. This knowledge makes possible the pictorial representation of chemical structures which are to be found throughout the literature of teaching and research in chemistry and biochemistry; typical chemical examples are shown in Fig. 1.1. Knowledge of the structure may be sought simply as a means of identifying a newly synthesized compound and understanding how it was formed, or the detailed geometry may be important for further investigations of reactivity, bonding, chirality, structure-energy relationships, etc.

Many experimental methods of probing the structure of a material are based on its absorption or emission of radiation; these are various forms of **spectroscopy**. Absorption takes place when the **frequency** ν of the radiation, and hence its quantum energy $h\nu$, matches a difference in certain energy levels in the sample. The observed frequencies of absorption thus provide information about energy levels and, from this information, something can be deduced about the structure of the material, based on a substantial body of accumulated experience.

For example, chemical shifts and coupling constants in proton NMR spectroscopy, obtained from the measured absorption spectrum, may indicate the number, chemical type, and relative proximity of the hydrogen atoms in a molecule and thus provide information on the connectivity (which atoms are bonded together); by more detailed NMR experiments some interatomic distances can be calculated. Similarly, the presence of particular functional groups in a molecule can be deduced from the appearance of characteristic absorption bands in an infrared (vibrational) spectrum.

In most spectroscopic techniques, what is measured is the variation of intensity of radiation passing through the sample as its frequency (or **wavelength**) is varied, in a particular direction; the intensity variation is produced by absorption of particular frequencies, leading to energy changes in the sample. This book is concerned with some diffraction methods, which are based on a different interaction of radiation with matter, usually in the solid state. Here we normally keep the wavelength fixed and measure the variation of intensity with direction, i.e. the scattering of **monochromatic** radiation is measured. From these measurements it is possible to work out the positions of the atoms in the sample and hence obtain a complete geometrical description of the structure. The intensity variation is caused by interference effects, also known as **diffraction**.

Monochromatic, literally 'single coloured', means having a single wavelength.

Spectroscopic and diffraction methods are thus based on different interactions of radiation with a sample, and provide complementary structural information; a thorough characterization of a material will often involve using both types of experiment. Diffraction methods are capable of providing much more detailed structural information than spectroscopic methods. There are, however, limitations on the types of materials which can be studied, as we shall see.

Diffraction effects are a characteristic behaviour of waves, including X-rays, light and other forms of **electromagnetic radiation**, involving interference effects as described later. An example can easily be seen by looking at a yellow street lamp (monochromatic light) through a finely woven fabric such as an umbrella. Simple diffraction patterns can also be produced on a screen or wall by shining a simple laser pointer through some clothing materials such as a cotton shirt.

A substantial body of rather complex mathematics forms the theory of diffraction methods for crystal structure determination. Fortunately it is not necessary to master this in order to understand the principles and application of the subject. This is, to a large extent, true even for those who carry out research in crystallography, because almost all the calculations are usually performed by sophisticated and largely automated computer programs. Advances in the subject since its birth just over a century ago have been very much in parallel with developments in computing, and modern personal computers are of sufficient power to make most **crystal structure determinations** very much faster than the days or weeks suggested by even fairly recent textbooks. Some of the other modern technological developments which have contributed to this dramatic increase in speed are described later in this book. The level of mathematics has deliberately been kept relatively low in this treatment of the subject. The fundamental equations for the diffraction process are given for completeness and to satisfy the more inquisitive reader, but they are also illustrated with analogies in words and diagrams in order to clarify their meaning.

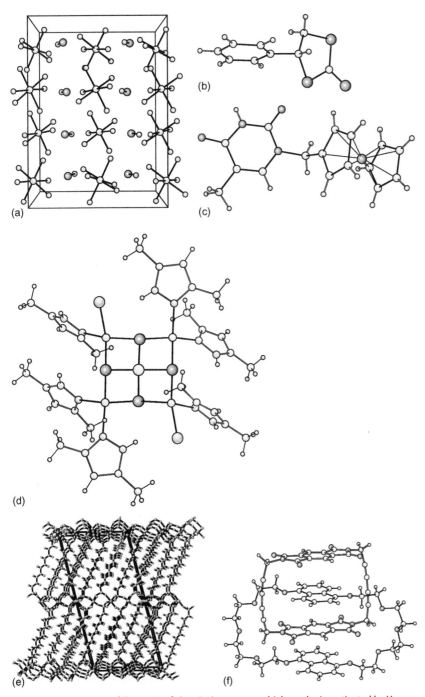

(a)

(b)

(c)

(d)

(e)

(f)

These structures have all been published: (a) Crystal structures of thionitrosyl hexafluoroantimonate(V) and thionitrosyl undecafluorodiantimonate(V) at 293 K and of thionitrosyl undecafluorodiantimonate(V) at 121.5 K: the effect of thermal motion on the apparent NS bond length. W. Clegg, O. Glemser, K. Harms, G. Hartmann, R. Mews, M. Noltemeyer and G. M. Sheldrick, *Acta Crystallogr. Sect. B* 1981, **37**, 548–552 (not in the CSD); (b) A bimetallic aluminum(salen) complex for the synthesis of 1,3-oxathiolane-2-thiones and 1,3-dithiolane-2-thiones. W. Clegg, R. W. Harrington, M. North and P. Villuendas, *J. Org. Chem.* 2010, **75**, 6201–6207 (CSD UCUDAZ); (c) Synthesis, structure and redox properties of ferrocenylmethylnucleobases. A. Houlton, C. J. Isaac, A. E. Gibson, B. R. Horrocks, W. Clegg and M. R. J. Elsegood, *J. Chem. Soc. Dalton Trans.* 1999, 3229–3234 (CSD BISLAQ); (d) Synthesis, crystal structures and spectroscopic characterization of two neutral heterobimetallic clusters $MS_4Cu_4(pz^{Me2})_6Cl_2$ (where M = Mo (**1**) or W (**2**), X = Cl (**1**) or disordered Cl/Br (**2**), and pz^{Me2} = 3,5-dimethylpyrazole). A. Beheshti, N. R. Brooks, W. Clegg and S. E. Sichani, *Polyhedron* 2004, **23**, 3143–3146 (CSD QALXOR); (e) Structural variety within gallium diphosphonates affected by the organic linker length. M. P. Attfield, Z. Yuan, H. G. Harvey and W. Clegg, *Inorg. Chem.* 2010, **49**, 2656–2666 (CSD YUSQIN); (f) Neutral [2]catenanes from oxidative coupling of π-stacked components. D. G. Hamilton, J. K. M. Sanders, J. E. Davies, W. Clegg and S. J. Teat, *Chem. Commun.* 1997, 897–898.

Fig. 1.1 An illustration of the range of chemical structures which can be investigated by X-ray crystallography: (a) the inorganic salt $[NS]^+[AsF_6]^-$; (b) a small chiral organic molecule; (c) a relatively small organometallic complex; (d) a polynuclear metal complex; (e) a polymeric network structure with organic pillars linking inorganic two-dimensional sheets; (f) a **supramolecular** assembly of two interlocking organic ring compounds.

Crystal structure determination can be applied to a wide range of size of structures, from very small molecules and simple salts to synthetic and natural polymers and to biological macromolecules such as proteins (Fig. 1.2). This book is concerned mainly with chemical applications, but some indication is given of the differences encountered when working with larger scale biological systems, particularly in Chapter 4.

One form of hen egg-white lysozyme was only the second protein crystal structure (and the first for an enzyme) to be determined by X-ray crystallography, in 1965.

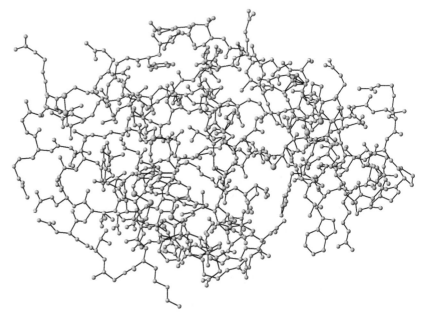

Fig. 1.2 A conventional ball-and-stick representation of one structural form of lysozyme, a relatively small protein; hydrogen atoms are omitted. The molecule contains C, N, O and S atoms.

1.3 The eye and microscope analogy

Objects of macroscopic size are visible to us because they scatter light falling on them. Our eyes intercept some of the scattered light and the function of the eye lens is to bring together the bundle of light rays, recombining the individual rays into an image on the retina (Fig. 1.3). Light consists of waves, and each scattered light ray has a particular intensity and a particular **phase** relative to other scattered rays, resulting from the scattering which produced it (Fig. 1.4). These relative intensities and phases, in turn, determine the nature of the image formed in the eye, which is understood by the brain as a representation of the object being viewed. Thus information on the shape (structure) of the object is carried in the intensities and phases of the light waves scattered by it. Since objects with different shapes can be distinguished simply by looking at them, it follows that they must have different, individual scattering patterns.

Visible light also has a range of wavelengths λ and frequencies ν, such that $\lambda\nu = c$, the velocity of light; for simplicity here and for comparison with the X-ray diffraction experiment, we consider light of just one wavelength, i.e. monochromatic light.

For smaller objects the eye needs help from more powerful lenses which can produce a larger image. The principle of operation is just the same: a

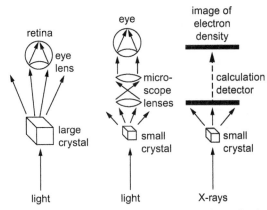

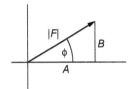

Fig. 1.4 **Amplitude** and **phase** of a wave; the phase becomes important only when two or more waves meet and combine.

Fig. 1.3 Left and centre, the function of the eye and microscope lenses for the recombination of scattered light; right, the equivalent two-stage process in X-ray diffraction.

proportion of the scattered light is collected by the **microscope** lens system and is **refracted** (the directions of the light rays are changed) to bring it all back together (the individual rays are combined by addition, with the correct relative amplitudes and phases) in the observer's eye (Fig. 1.3). Note that it is not necessary to capture all the scattered light, but the observed image becomes less clear as the proportion of light collected is reduced: a good quality image is produced by a wide objective lens close to the object.

The lower limit on the size of objects which can be clearly seen with a microscope of sufficient magnifying power is set, not by optical engineering capabilities, but by the wavelength of visible light itself (in the range 400–700 nm). Objects which are much smaller than this, such as individual molecules (which are typically 100–1000 times smaller), do not give any significant scattering of the light. In order to 'see' the structure of molecules, it is necessary to resolve the component atoms, which are of the order of one or a few angstroms in size. Instead of visible light, this means using X-rays. So, in principle, what we need is an X-ray microscope in order to observe molecular structure.

Quite apart from safety issues and the need to provide suitably sensitive detectors to record the X-rays, such an instrument does not exist, because conventional lenses cannot be used to focus X-rays. The scattering of X-rays by molecules does, indeed, occur, but the scattered X-rays cannot be physically recombined to form an image.

The situation is, however, not hopeless, because the pattern of scattered X-rays can be directly recorded, either on photographic film (now mainly of historical interest) or on a variety of other X-ray sensitive detectors, then the recombination which is impossible physically can be performed mathematically, with the aid of computers: the mathematics involved is well established, but it requires a considerable amount of calculation. Thus the experiment to determine a molecular structure falls into two parts, recording the X-ray scattering pattern and carrying out the recombination subsequently by mathematics (Fig. 1.3), and is no longer instantaneous like viewing an object through a microscope.

Refraction is the alteration in the direction of travel of light as it passes from one medium into another with a different **refractive index**. It is responsible, for example, for the apparent bending of a drinking straw in a glass of water (and many other optical illusions), because water and air have quite different refractive indices. Refraction should not be confused with diffraction, a quite different phenomenon despite the similar name; even senior chemistry researchers sometimes produce nonsense words such as 'defraction'.

The **angstrom unit**, Å, is not strictly permitted by the SI rules, but is widely used in structural chemistry because of its convenient size: 1Å = 100 pm = 0.1 nm.

Focusing of extremely high intensity X-rays can be achieved using special methods, but they are not of general application.

Note that it is perfectly possible to record the pattern of visible light scattered by an object in an analogous way; some examples will be used in Section 1.5 to illustrate the principles of diffraction. Although this is not commonly done, a variant of it is a valuable procedure used by mineralogists, in identifying and characterizing mineral specimens.

The technique is known as **crystal structure determination** because the object studied is actually a small crystalline sample rather than a single molecule, which it would be impossible to hold in the X-ray beam for the duration of the experiment and which would give an immeasurably weak scattering pattern on its own. In a crystal, there are large numbers of identical molecules (or molecules and their mirror images), locked in position in a regular arrangement, which together give significant scattering. The method can be used only for samples which can be obtained in a suitable crystalline form.

When the method is successful, it provides an image of the molecular structure. More precisely, it locates the components of the material which interact with the incident X-rays and scatter them. These are the electrons in the atoms. Although each individual electron/X-ray interaction is instantaneous, the total time taken to record the scattering pattern with modern equipment is usually minutes or hours, and even the most rapid methods available are very slow compared with the movement of electrons, so the picture that results is of a time-averaged electron density (Fig. 1.5). Concentrations of electron density in the image correspond to atoms, somewhat spread out by time-averaged vibration, and the results are usually interpreted and presented as atomic positions, but there are some important consequences of the fact that the primary result is the location of electron density, which will be discussed later.

One very important consequence of the need to divide the overall experiment into two parts instead of directly recombining the scattered X-rays to generate an image is that some of the information in the scattered X-rays is unfortunately lost. When the X-ray scattering pattern is recorded, the individual wave amplitudes are retained as relative intensities (intensity is proportional to the square of amplitude), but the relative phases are lost. This makes the mathematical reconstruction stage much less straightforward. It is one of the fundamental challenges of crystallography and methods of dealing with 'the phase problem' have been major research projects throughout the history of the subject.

$I \propto |F^2|$

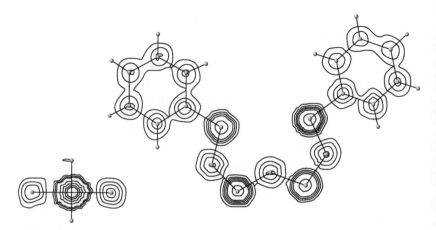

Fig. 1.5 An electron density map, with the positions of atoms and bonds marked. Note that the rather wiggly contours here and in other figures are an artefact of the computer program used.

1.4 Fundamentals of the crystalline state

A perfect crystalline solid material is made up of a large number of identical molecules which are arranged in a precisely regular way repeated in all directions, to give a highly ordered structure. Even for a microscopic crystal, the repetition is effectively infinite on an atomic scale. This repetition of a structural unit by pure **translation**, to form a space-filling, three-dimensional crystal, is a type of symmetry, which occurs in all crystalline solids, whether or not they also show other forms of symmetry such as rotation, reflection or inversion.

The basic unit of a crystal structure may not be a single molecule, but a number of ions, an assembly of a few molecules, or other unit, which is then repeated. Real structures also show various kinds of defects and irregularities, which are beyond the scope of this book.

In two dimensions, such translation symmetry is familiar in the form of patterns on wallpapers, flooring and other manufactured materials. A two-dimensional projection of part of a real crystal structure is shown in Fig. 1.6(a). The basic structural unit here is a single molecule. All the molecules are identical and repetition by translation gives the complete two-dimensional pattern; the structure has no symmetry other than pure translation.

If each molecule is represented by just a single point (placed, for example, on the same atom in each molecule), the result is a regular array of points, which shows the repeating nature of the structure but not the actual form (the detailed contents) of the basic structural unit (Fig. 1.6(b)). This array of identical points, equivalent to each other by translation symmetry, is called the **lattice** of the structure.

To define the repeat geometry of the structure, a parallelogram of four lattice points is chosen, and is called the **unit cell** of the structure; it has two sides of different lengths and one included angle (Figs 1.6(c) and 1.7). Obviously, many different choices of unit cell are possible for any one lattice (Fig. 1.6(d)); there are conventions to guide this choice. In the absence of any rotation or reflection symmetry in the structure, the conventional unit cell has sides as short as possible, $a \leq b$, and an angle as close as possible to 90°. Inspection shows that each unit cell in Fig. 1.6(c) contains parts of several molecules, and the total contents are just one molecule. Each unit cell contains the equivalent of one lattice point (one repeat unit).

The presence of rotation or reflection symmetry in a crystal structure, relating molecules or parts of molecules to each other, imposes restrictions on the geometry of the lattice and unit cell. For example, fourfold rotation symmetry in a two-dimensional lattice requires a square unit cell with two equal sides and a 90° angle, while reflection symmetry gives a 90° angle but still allows the two unit cell sides to be of different length (Fig. 1.8).

For an introduction to symmetry in chemistry, which is largely concerned with molecular symmetry and hence with point groups, reference should be made to standard chemistry text books.

In three dimensions, a unit cell has three sides and three angles (Fig. 1.9). Conventionally, the three lengths are called a, b, c, and the angles α, β, γ such that α lies between the b and c axes, i.e. opposite a. In the absence of any rotation or reflection symmetry, the three axes have different lengths and the three angles are different from each other and from 90°. Rotation and reflection symmetry impose restrictions and special values on the unit cell parameters. On the basis of these restrictions, crystal symmetry is broadly divided into seven types, called the seven **crystal systems**. Table 1.1 shows their names, minimum symmetry

Unit cell parameters are also known as **lattice parameters**.

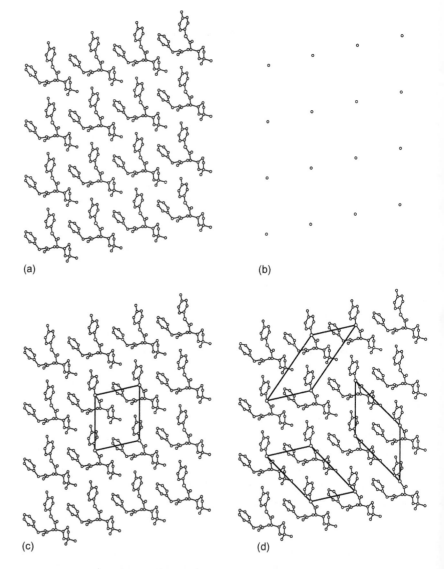

(a)   (b)

(c)   (d)

Fig. 1.6 (a) A two-dimensional projection of an organic crystal structure, showing 16 identical molecules; (b) the two-dimensional lattice for this pattern; (c) a unit cell for the pattern; (d) other possible choices of unit cell for the same pattern.

Fig. 1.7 The geometry of the unit cell for Fig. 1.6.

characteristics and unit cell geometries. Note that inversion symmetry does not impose any lattice geometry restrictions, since every three-dimensional lattice has inversion symmetry anyway, whether the complete structure represented by the lattice is centrosymmetric or not. As we shall see later, this has important consequences in X-ray diffraction.

Individual objects, such as molecules, can display rotation symmetry of any order C_2, C_3, ... up to C_∞ rotation axes, but only C_2, C_3, C_4, and C_6 axes can be found in true crystals. This does not mean that molecules with C_5 symmetry, for

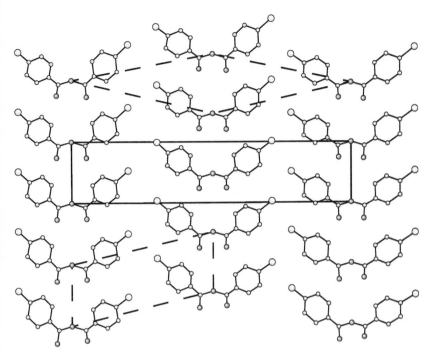

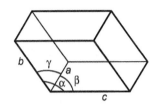

Fig. 1.9 A three-dimensional unit cell.

Fig. 1.8 A structure with reflection symmetry. The reflection lines run vertically along the unit cell edges and through the cell centre, bisecting each molecule. There are also glide reflection lines half-way between these (see later). The conventional centred rectangular unit cell is outlined, together with two possible primitive cells, each of half the area.

example, (such as ferrocene $(C_5H_5)_2Fe$, or buckminsterfullerene C_{60}) cannot form crystalline solids, but the rotation symmetry does not apply to the surroundings of the molecule and to the structure as a whole (Fig. 1.10).

On the other hand, crystals can have other types of symmetry element not possible in single finite molecules, in which rotation or reflection is combined

Table 1.1 Crystal systems

Crystal system	Essential symmetry	Restrictions on unit cell
Triclinic	none	none
Monoclinic	one twofold rotation and/ or mirror	$\alpha = \gamma = 90°$
Orthorhombic	three twofold rotations and/ or mirrors	$\alpha = \beta = \gamma = 90°$
Tetragonal	one fourfold rotation	$a = b; \alpha = \beta = \gamma = 90°$
Trigonal	one threefold rotation	$a = b; \alpha = \beta = 90°; \gamma = 120°$
Hexagonal	one sixfold rotation	$a = b; \alpha = \beta = 90°; \gamma = 120°$
Cubic	four threefold rotation axes	$a = b = c; \alpha = \beta = \gamma = 90°$

Although trigonal and hexagonal crystals are characterized by the same shape of unit cell, they have different essential symmetries (threefold and sixfold rotations, respectively); the unit cell shape alone does not distinguish them. Note also that some structures have unit cells that approximate to the shape of a higher-symmetry system, e.g. it is possible for the angle β to be insignificantly different from 90° for a monoclinic structure, so that the unit cell appears to be of orthorhombic symmetry; this can complicate the determination of the correct symmetry and can lead to the phenomenon of twinning, discussed later.

It is often the case that molecules occupy positions in crystal structures with lower symmetry than their own intrinsic point-group symmetry. This can have consequences, not only in crystallography, but also in spectroscopy and in physical properties; for example, bands in solid-state spectra often show splittings compared with those from samples in solution, because atoms become non-equivalent by symmetry in the solid.

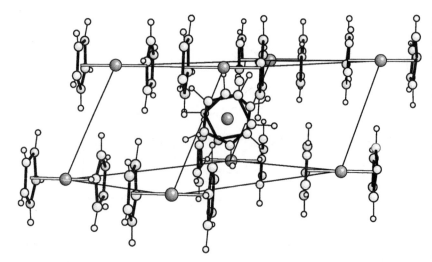

Fig. 1.10 The structure of one crystalline form of ferrocene, showing that the surroundings of each molecule do not have fivefold rotation symmetry.

with translation to give, respectively, **screw axes** and **glide planes**. Figure 1.11 illustrates the difference between simple reflection and glide reflection in a two-dimensional pattern. Reflection symmetry is familiar in everyday life; the two mirror-related objects lie directly opposite each other, reflected from each other across the mirror line (in two dimensions) or mirror plane (in three dimensions). Glide reflection involves displacement of the two mirror images relative to each other by exactly half of a repeat unit of the pattern. In a two-dimensional pattern such as in Fig. 1.11, there is only one possibility for the direction of glide, parallel to the glide line. In a three-dimensional crystal structure, the direction

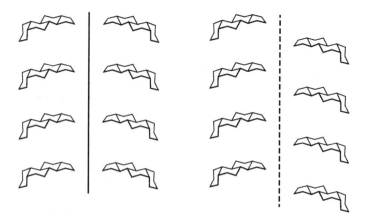

Fig. 1.11 A two-dimensional illustration of simple reflection (left) and glide reflection (right) for a regularly spaced repeated motif. Because glide reflection provides more efficient packing of molecules than does simple reflection, space groups with glide planes tend to be observed more often than those with mirror planes. A similar argument applies to screw axes versus simple rotation axes.

of glide can, in most cases, be parallel to either of two different axes or along the diagonal between them. Similarly, screw axes combine a simple rotation with a translation along the direction of the axis. An example of glide reflection symmetry occurring in the same structure as simple reflection symmetry can be seen in Fig. 1.8.

For some structures with more than purely translation symmetry, it is convenient and conventional to choose a unit cell which contains more than one lattice point, because the resultant unit cell geometry displays the symmetry more clearly. The unit cell in Fig. 1.8 has a lattice point at each corner, but also an entirely equivalent lattice point at its centre; this **centred unit cell** is a rectangle, with a 90° angle, whereas a **primitive unit cell** with lattice points only at its corners and with half the area, would not be. A three-dimensional example is the all-face-centred cubic structure of many metals, in which the centre of each face of the unit cell is equivalent to all the cell corners (Fig. 1.12). This unit cell has four times the volume of the smallest possible primitive unit cell, but has the advantage of displaying the cubic symmetry clearly in its shape.

Symmetry elements in a single molecule all pass through one point, and the various possible combinations of symmetry elements are known as **point groups**. In a crystal, symmetry elements do not all pass through one point, but they are regularly arranged in space in accordance with the lattice translation symmetry. There are exactly 230 possible arrangements of symmetry elements in the solid state; these are called the 230 **space groups**, and their symmetry properties are well established and available in standard reference books and tables, the most comprehensive and widely used being the **International Tables for Crystallography**, Volume A.

Some popular misconceptions about lattices and unit cells should be dispelled.

- The term 'lattice' is often used as a synonym for 'structure', but this is incorrect, because the lattice shows only the repeating nature of the structure, not the detailed contents. A particularly common misuse of the term is 'lattice solvent' instead of 'uncoordinated solvent'.

- *Any* point in a crystal structure can be chosen as a lattice point and the lattice constructed from all the equivalent points (with identical environments in identical orientation); conventions apply to the choice of lattice points

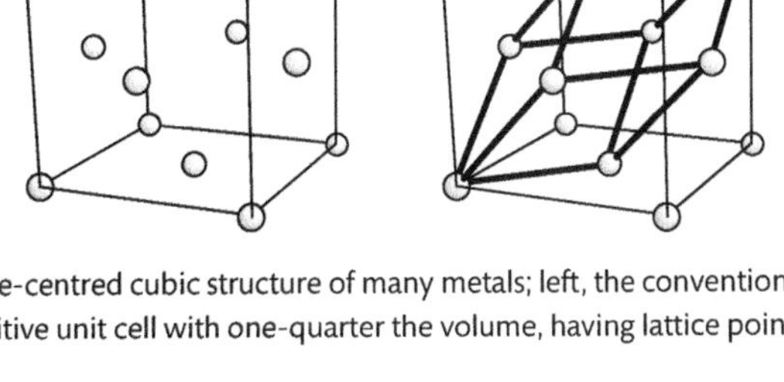

A primitive unit cell has lattice points only at its eight corners; all of these, by definition of a lattice, are entirely equivalent. A centred unit cell has other, also entirely equivalent, lattice points. For a two-dimensional lattice there is only one form of centring, with an equivalent lattice point exactly at the centre of each unit cell. For a three-dimensional lattice, various different types of centring are possible, with lattice points at the centres of pairs of opposite faces, at the centres of all faces, or at the very centre (the body centre) of each unit cell.

A point group may be thought of as the complete collection of all symmetry elements passing through a central point, describing the symmetry of an individual object. A space group is the complete collection of all symmetry elements for an infinitely repeating pattern. Both can be elegantly treated by mathematical group theory; strictly speaking, for this to be true, point groups and space groups are actually complete collections of symmetry operations rather than symmetry elements, but the distinction is not important for our purposes here.

The *International Tables for Crystallography* are published for the International Union of Crystallography by John Wiley & Sons. They are available as an online resource at http://it.iucr.org/ to subscribing institutions.

Fig. 1.12 The face-centred cubic structure of many metals; left, the conventional cubic unit cell; right, a primitive unit cell with one-quarter the volume, having lattice points only at its corners.

relative to the positions of symmetry elements in the structure (in particular, it is conventional to place lattice points on inversion centres when they are present), and it is the exception rather than the rule for an atom to lie on a lattice point in any other than simple high-symmetry structures (which are usually the structures first encountered by students in the context of lattices and unit cells).

* The unit cell is the repeat unit (building block) of any crystal structure and so contains a small whole number of molecules, but in most structures the molecules lie across unit cell edges and faces rather than being neatly contained within these purely mathematical constructions (see, for example, Figs 1.6 and 1.8).

Pure lattice translation symmetry relates individual unit cells to each other. If a structure has any other symmetry as well, then this symmetry relates atoms and molecules within one unit cell to each other. Thus the unique, independent part of the structure is usually only a fraction of the unit cell, the fraction depending on the amount of symmetry present. This unique portion is called the **asymmetric unit** of the structure. Operation of all the rotation, reflection, inversion, and translation symmetry elements of the space group on this asymmetric unit generates the complete crystal structure. The asymmetric unit may consist of one molecule, more than one molecule, or a fraction of a molecule, the molecule itself possessing symmetry that is displayed by the crystal structure as a whole.

One further point should be made about symmetry. Different symbols are used for symmetry elements, and for combinations of them (point groups and space groups) in different fields of science. For rotation, reflection, and inversion symmetry elements, which can occur both in individual molecules and in solid-state structures, the correspondence of the two common sets of symbols is shown in Table 1.2. There are good reasons for the differences in the symbols, and also for the different definitions of the so-called 'improper rotations', but they do lead to confusion. The Schoenflies notation, used in molecular spectroscopy, produces convenient and compact symbols for point groups, while

A *proper rotation* is a simple rotation axis. An *improper rotation* combines the operation of rotation with either inversion through a point (crystallography) or reflection in a perpendicular plane (spectroscopy).

Table 1.2 Correspondence of symbols for symmetry elements in the Hermann–Mauguin (crystallography) and Schoenflies (spectroscopy) systems of notation.

	Crystallography	Spectroscopy
Proper rotations	2	C_2
	3	C_3
	4	C_4
	6	C_6
Improper rotations	$\bar{3}$	S_6
	$\bar{4}$	S_4
	$\bar{6}$	S_3
Reflection	m	σ
Inversion	$\bar{1}$	i

the Hermann–Mauguin notation (also known as international notation), used in crystallography, is much better suited to space group representation, some of which will be seen later in examples. The details of symbols for glide planes and screw axes are beyond the scope of this short text.

1.5 Diffraction of X-rays by molecules and crystals

Figure 1.13 shows part of the pattern of scattered X-rays produced by a single crystal. The complete pattern can only be recorded by rotating the crystal in the X-ray beam, for reasons we shall see later. There are many different kinds of instrument for recording X-rays scattered by crystals, and they produce a variety of appearances, but in each case a good quality crystal always gives a pattern of spots of varied intensity. In Fig. 1.13, different intensities are represented by different sizes of spot; on a photographic film, they would vary in their degree of blackness.

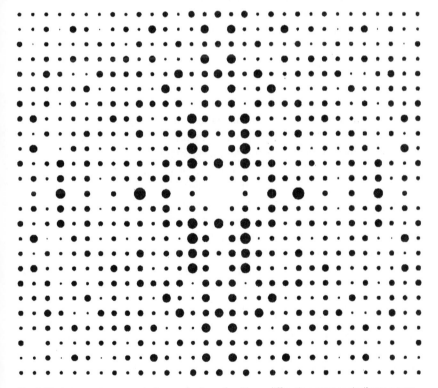

Fig. 1.13 A computer-generated reproduction of an X-ray diffraction pattern similar to a type obtained by one form of X-ray camera.

The pattern of spots has three properties of interest, which correspond to three properties of the crystal structure.

- First, the pattern has a particular *geometry*. The spots lie in certain positions which are clearly not random. Each spot is generated at the detector by an individual scattered X-ray beam travelling in a definite direction from the crystal. This diffraction pattern geometry is related to the lattice and unit cell geometry of the crystal structure and so can tell us the repeat distances (and directions) between molecules.

- Second, the pattern has *symmetry*, not only in the regular spatial arrangement of the spots but also in having equal intensities of spots which lie in symmetry-related positions relative to the centre of the pattern. The diffraction pattern symmetry is closely related to the symmetry of the unit cell of the crystal structure, i.e. to the crystal system and space group. The pattern in Fig. 1.13 has both vertical and horizontal reflection symmetry, and an inversion point at the centre.

- Third, apart from this symmetry, there is no apparent relationship among the *intensities* of the individual spots, which vary widely; some are very intense, while others are too weak to be seen (their positions are deduced from the regular array). These intensities hold all the available information about the positions of the atoms in the unit cell of the crystal structure, because it is the relative atomic positions which, through the combination of their individual interactions with the X-rays, generate different amplitudes for different directions of scattering.

Thus, measurement of the geometry and symmetry of an X-ray scattering pattern provides information on the unit cell geometry and symmetry, while determination of the full molecular structure involves the measurement of all the many individual intensities, a considerably greater task.

An understanding of these relationships is valuable in grasping the fundamentals of X-ray diffraction in crystal structure determination. They can be summarized elegantly in mathematical terms, and the main equations will be presented later, but a pictorial approach may be more informative to many readers and will be given first. To illustrate the relationships between objects and their scattering patterns simply, we restrict the treatment to two-dimensional objects and scale up the process by a factor of about 10^4, so that atoms in planar molecules are represented by small holes punched in opaque card and monochromatic X-rays are replaced by monochromatic light with a wavelength comparable to the hole size and spacing. By doing this, it is possible to obtain from a modified type of microscope both an image (a picture) of each object and its light scattering pattern (also known as its optical transform) for comparison. Twelve pairs of objects and corresponding optical transforms are shown in Fig. 1.14.

Note first that every object produces a different pattern. These patterns of light and dark are generated by interference effects, i.e. by the combination of light waves coming from different parts of the object as the incident light passes through it.

Even for a single circular hole (think of this as a single atom in an X-ray beam) there are interference effects for the light waves scattered by the edges of the hole (objects **A** and **B**). In some directions these waves are in phase and the scattering pattern is bright; in other directions, scattered waves are out of phase and cancel each other, so little or no net intensity is seen. Thus the transmitted light does not give a sharp pattern matching the shape and size of the hole, but a diffuse pattern with a central circular bright region surrounded by rings of lower intensity. A larger hole (**B**) gives a smaller pattern. A rectangular hole (**C**) gives a diffuse pattern of rectangular symmetry, also with light and dark fringes. Note that the relative dimensions of the rectangle (tall and narrow) are reversed in the scattering pattern (short and wide). A series of experiments with different rectangles shows that there is an exact inverse relationship.

Several holes together represent a single molecule with a number of atoms (**D** to **F**). These give more complicated scattering patterns, each with diffuse areas of varied intensity. There are now additional scattered wave interference effects, which depend on the relative positions of the different holes. The rectangular object (**D**) gives a pattern with rectangular symmetry. The regular hexagon of holes (**F**) gives a pattern with the same sixfold symmetry. The object with only one vertical reflection line of symmetry (**E**), however, generates a pattern with extra symmetry: two mutually perpendicular lines of reflection, horizontal and vertical, intersecting in an inversion point. In general, each optical transform has the same symmetry as the corresponding object, with the addition of inversion symmetry if it is not already present (and this may imply further symmetry elements as in this case); the scattering pattern never has less symmetry than the object. In three dimensions, an equivalent rule applies, with the addition of an inversion centre to all scattering patterns.

Parts **G** to **I** show the effect of pure translation symmetry on the scattering of radiation by an object. Again, extra interference effects take place for the light rays scattered from individual holes. Because the holes are regularly spaced, these interference effects strongly reinforce each other, and the most obvious result is that the scattering patterns contain sharp maxima of intensity instead of broad diffuse regions. A single row of holes gives a pattern of narrow bright stripes running perpendicular to the row (**G** and **H**). A two-dimensional regular array of holes (**I**) imposes a restriction on the intensity maxima in both dimensions simultaneously, so that only sharp points of light are now seen (where two perpendicular sets of stripes cross), and these are also regularly spaced. The rows of spots in the optical transform always lie perpendicular to the rows of holes in the object, and there is an inverse relationship in the spacings: the array of holes in the object **I** are spaced wider horizontally than vertically, and the bright spots in the scattering pattern are spaced wider vertically than horizontally, with the inverse ratio.

This interference effect is well known and exploited in many branches of science, and is called **diffraction**. Here we see that a regular lattice arrangement of objects scattering radiation produces severe diffraction restrictions, so that the scattered radiation has significant intensity only in certain well-defined directions and not in a diffuse pattern such as occurs with a single object.

Finally, we place more complicated objects (hexagons, representing molecules, instead of single holes) on lattices. The three lattices **J** to **L** have different geometries (different unit cells) but each has as its basic repeat unit the same single hexagon in the same orientation (which is the same as **F**, but turned through 90°). Each

The extra bright fringes and spots near the centres of these patterns are a result of the small number of holes in the objects. With more holes, these subsidiary maxima decrease in intensity and eventually disappear, and the main maxima become sharper. A real crystal may contain millions of unit cells in each direction, so the maxima for scattered X-rays are sharp.

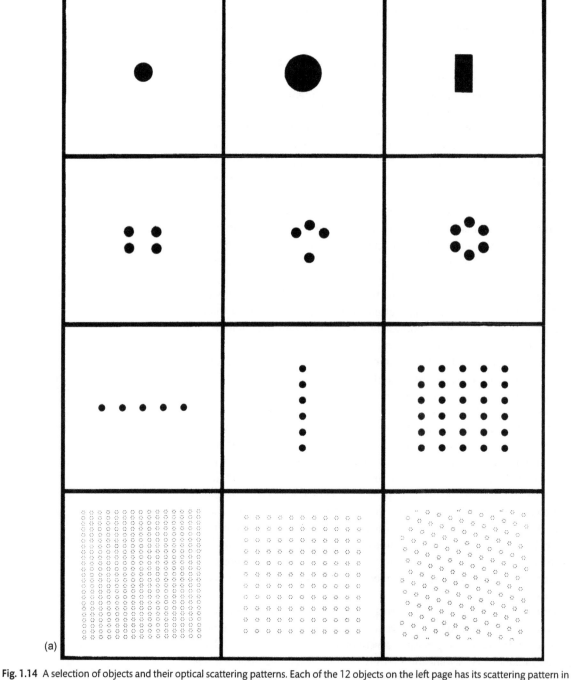

(a)

Fig. 1.14 A selection of objects and their optical scattering patterns. Each of the 12 objects on the left page has its scattering pattern in the corresponding position on the right page. The bottom row of objects is on a different scale; in the objects themselves, each hexagon has the same size as the individual hexagon in the second row. In the description in the text, the objects and their patterns are labelled from (**A**) to (**C**) in the top row, continuing to (**L**) at the end of the bottom row. These illustrations are taken from a very comprehensive compilation in *Atlas of Optical Transforms* by G. Harburn, C. A. Taylor and T. R. Welberry (G. Bell and Sons Ltd, London, 1975), with the permission of the authors.

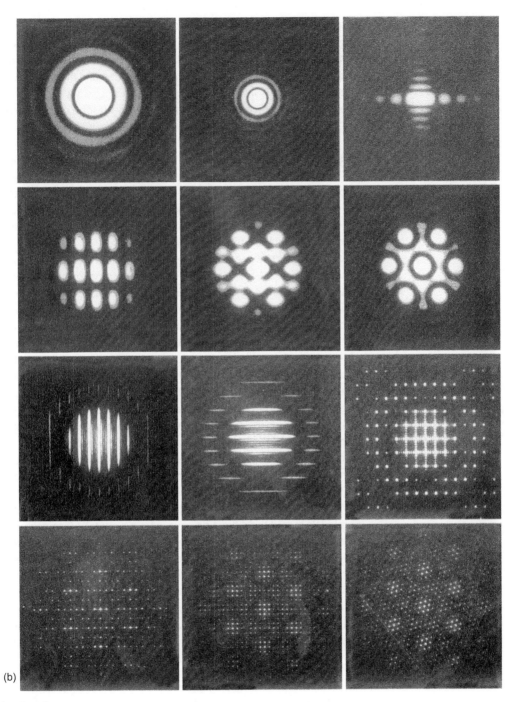

(b)

Fig. 1.14 (*continued*)

produces a diffraction pattern consisting of a regularly spaced array of more or less bright points. The positions of these points are dictated by the diffraction conditions, generated by the parent lattice: in each case, the rows of points run perpendicular to the rows of hexagons on the object lattice, and the spacing of the points in each direction is inversely proportional to the spacing of the hexagons. The intensities of the individual spots are produced by the form of the single object (the hexagon): comparison of **J**, **K**, and **L** with **F** (especially with half-closed eyes to blur the lattice diffraction effects!) shows that the underlying pattern of light and dark is the same for all of them, and this is the optical transform of the single hexagon.

The net effect is like looking at the optical transform of the single hexagon through a sieve. The transform itself (the variation of intensity) is determined by the detailed geometrical form of the single object, while the mesh of the sieve, which dictates the points at which the transform intensities can actually be seen, is determined by the lattice geometry.

Extending this to three dimensions, translating it to the case of a single crystal in a monochromatic X-ray beam, and introducing some formal terminology, these relationships illustrated pictorially are summarized as follows. An object scatters radiation of wavelength comparable to its own size; the mathematical relationship between the object and the scattering pattern is *Fourier transformation*, such that the scattering pattern is the **Fourier transform** of the object, and the image of the object (provided by an optical microscope or by X-ray crystallography) is in turn the Fourier transform of the scattering pattern. If identical objects are arranged on a lattice, diffraction effects of the lattice are also imposed, so that the diffraction pattern can have non-zero intensity only where the direction of scattering satisfies the equations for diffraction geometry. The overall effect is a combination of the two effects—scattering by the object further restricted by diffraction by the lattice—so the observed diffraction pattern is the Fourier transform of the single object sampled at certain geometrically determined points.

Fourier transformation is a well-known and well-established mathematical operation used in a wide variety of science and technology applications, including spectroscopy and image processing.

The previous paragraph is the essential basis of the technique of X-ray crystallography. In principle, then, the process of crystal structure determination is simple. We record the diffraction pattern from a crystal. Measurement of the diffraction pattern geometry and symmetry tells us the unit cell geometry and gives some information about the symmetry of arrangement of the molecules in the unit cell. Then from the individual intensities of the diffraction pattern we work out the positions of the atoms in the unit cell, by pretending to be a microscope lens system, adding together the individual waves with their correct relative amplitudes and phases. And here we see the **phase problem**, the fact that the measured diffraction pattern provides directly only the amplitudes and not the required phases, without which the Fourier transformation cannot be made.

1.6 The geometry and symmetry of X-ray diffraction

Geometry

Having seen the fundamental basis of X-ray diffraction in both pictorial and verbal form, we will now present the relationships mathematically.

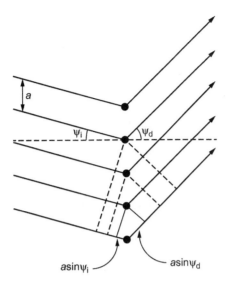

Fig. 1.15 Diffraction by a single row of regularly spaced objects.

For the geometry, consider first diffraction by a single row of regularly spaced points (one-dimensional diffraction; Fig. 1.15).

In any particular direction, the radiation scattered by the row of points will have zero intensity by destructive interference of the individual scattered rays unless they are all in phase. Since, except in the straight-through direction, individual rays have different path lengths, these path differences must be equal to whole numbers of wavelengths to keep the rays in phase. So, for rays scattered by two adjacent points in the row,

$$\text{path difference} = a\sin\psi_i + a\sin\psi_d = h\lambda \qquad (1.1)$$

where ψ_i and ψ_d are the angles of the incident and diffracted beams as shown, λ is the wavelength, a is the one-dimensional lattice spacing, and h is an integer (positive, zero, or negative). For a given value of ψ_i (a fixed incident beam), each value of h corresponds to one of the observed diffraction maxima and the equation can be used to calculate the permitted values of ψ_d, the directions in which intensity is observed. The result, as we have seen in Fig. 1.14, is a set of bright fringes.

For diffraction by a three-dimensional lattice there are three such equations and all have to be satisfied simultaneously. The first equation contains the lattice a spacing, angles relative to this a axis of the unit cell, and an integer h. The other two equations, correspondingly, contain the unit cell axes b and c, and integers k and l respectively.

Thus each allowed diffracted beam (each spot seen in an X-ray diffraction pattern) can be labelled by three integers, or **indices**, hkl, which uniquely specify it if the unit cell geometry is known.

These three equations for diffraction geometry, the **Laue conditions**, are cumbersome to use in this form. An alternative but equivalent description was

The letters h, k, and l are used conventionally by all crystallographers although, unlike other conventional triplets of letters used in the subject (such as a, b, c; x, y, z), they are not consecutive in the alphabet, for unimportant historical reasons.

At the age of 24 W. L. Bragg, together with his father W. H. Bragg, was awarded the Nobel Prize for Physics for this work and its application in the first crystal structure determinations in 1913. Max von Laue received the Prize the previous year for his part in the discovery of the diffraction of X-rays by crystals in 1912.

derived by W. L. Bragg soon after the experimental demonstration that X-rays could be diffracted by crystals, and is expressed in the single **Bragg equation**, which is universally used as the basis for X-ray diffraction geometry (Fig. 1.16). Bragg showed that every diffracted beam that can be produced by an appropriate orientation of a crystal in an X-ray beam can be regarded geometrically as if it were a reflection from sets of parallel planes passing through lattice points (**lattice planes**), analogous to the reflection of light by a mirror, in that the angles of incidence and reflection must be equal and that the incoming and outgoing beams and the normal to the reflecting planes must themselves all lie in one plane. The reflection by adjacent planes in the set gives interference effects equivalent to those of the Laue equations; to define a plane we need three integers to specify its orientation with respect to the three unit cell edges, and these are the indices *hkl*; the spacing between successive planes is determined by the lattice geometry, so is a function of the unit cell parameters.

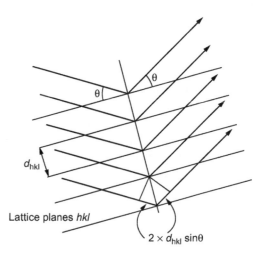

Fig. 1.16 The Bragg construction for diffraction by a three-dimensional crystal structure; one set of parallel **lattice planes** is seen edge-on.

For rays reflected by two adjacent planes,

$$\text{path difference} = 2d_{hkl} \sin\theta = n\lambda \tag{1.2}$$

In practice, the value of *n* can always be set to 1 by considering planes with smaller spacing (*n* = 2 for planes *hkl* is equivalent to *n* = 1 for planes 2*h*, 2*k*, 2*l* which have exactly half the spacing) and it is in the form:

$$\lambda = 2d_{hkl} \sin\theta \tag{1.3}$$

that the Bragg equation is always used. It allows each observed diffracted beam (commonly known as a 'reflection') to be uniquely labelled with its three indices and for its net scattering angle (2θ from the direct beam direction) to be calculated from the unit cell geometry, of which each d_{hkl} spacing is a function.

Rearrangement of the Bragg equation gives:

$$\sin\theta = \left(\frac{\lambda}{2}\right) \times \left(\frac{1}{d_{hkl}}\right) \qquad (1.4)$$

The distance of each spot from the centre of an X-ray diffraction pattern such as Fig. 1.13 is proportional to $\sin\theta$ and hence to $1/d_{hkl}$ for some set of lattice planes. This demonstrates mathematically the reciprocal (inverse) nature of the geometrical relationship between a crystal lattice and its diffraction pattern, already seen pictorially.

The Bragg equation is the basis of all methods for obtaining unit cell (lattice) geometry from the measured geometry of the diffraction pattern. The exact application depends on the experimental setup used to obtain the diffraction pattern.

Symmetry

The unit cell dimensions provide some information about the distances between molecules, which are regularly spaced in the crystal structure. In most structures, however, each unit cell contains not one but several molecules which are related to each other by the space group symmetry. This symmetry is revealed in various aspects of the appearance of the diffraction pattern, from which it is usually possible to choose the correct space group from the complete list of 230 or, at least, to narrow down the choice to a few possibilities; ambiguities arise, for example, because a diffraction pattern may have more symmetry than the structure itself, but it cannot have less, and in such cases the correct answer is known only when the structure is successfully solved and refined.

For a compound of known chemical formula, the number of molecules per unit cell can be calculated (if the density of the crystals is measured or estimated, as illustrated by examples that follow). This number can be compared with the number of asymmetric units required by the symmetry elements present in the space group (see Section 1.4). If the two are equal, then there is one molecule per asymmetric unit and this tells us nothing about the molecular shape. If, however, the asymmetric unit is only a fraction of a molecule, then the molecule itself must have one or more symmetry elements of the space group, and this provides some information on the molecular shape, even before the full structure determination is carried out.

This is best illustrated with examples. The first is an organic compound (an oxepin), the chemical structure of which is shown in Fig. 1.17. Crystals obtained from solution in toluene (methylbenzene, $C_6H_5CH_3$) belong to the monoclinic crystal system, with $a = 13.616$, $b = 14.295$, $c = 16.520$ Å, $\beta = 95.18°$; the unit cell volume $V = 3202.4\,\text{Å}^3$.

The density D of the crystals is 1.23 g cm^{-3}. Since the density and unit cell volume are known experimentally, the mass of the contents of one unit cell can be calculated.

$$\text{mass} = \text{density} \times \text{volume}$$
$$\text{so, unit cell mass} = 1.23\,\text{g cm}^{-3} \times 3202.4\,\text{Å}^3$$
$$= 1.23\,\text{g cm}^{-3} \times 3202.4 \times (10^{-8})^3\,\text{cm}^3$$
$$3.939 \times 10^{-21}\,\text{g per unit cell} \qquad (1.5)$$

A more detailed treatment of diffraction geometry, in the form of the **Ewald sphere**, can be found in other crystallography texts. It is of particular importance in the processing of data collected using modern area-detector diffractometers, but it is unnecessary here and the usual presentation in two dimensions of a three-dimensional construction can be misleading. Similarly, the most elegant description of diffraction in crystallography uses the concept of the **reciprocal lattice**, but this requires familiarity with **vector** algebra, which is not assumed of the readership of this book, and goes beyond our needs.

Simple rotation and reflection symmetry is seen directly in the diffraction pattern, always with the addition of an inversion centre if it is not already there. Glide planes and screw axes cause particular subsets of reflections to have exactly zero intensity, according to well-established rules for these **systematic absences**. The effect can be seen in the central rows, both horizontal and vertical, in Fig. 1.13, where alternate reflections have zero intensity.

For a unit cell with all angles equal to 90°, the volume is simply $V = abc$; if two of the cell angles are 90° and one is not (as in this monoclinic example, for which the non-orthogonal angle is β), then $V = abc \sin\beta$. This covers all except the triclinic system (and **rhombohedral** structures, which are a subset of the trigonal system in which the standard hexagonal-type unit cell has 3 rather than 1 lattice point and the corresponding primitive unit cell has $a = b = c$; $\alpha = \beta = \gamma$ ($\neq 90°$)–this may be thought of as a cube that is either compressed or elongated along one of its four body diagonals, the other three no longer showing threefold rotation symmetry); for the triclinic system the formula for the unit cell volume, involving all six cell parameters, is rather more complicated unless it is expressed in vector notation.

Be careful of the units in these calculations; they need to be consistent, and initial data usually have to be multiplied or divided by some powers of 10 to achieve this. For simplicity, uncertainties in the experimental measurements are ignored here; this topic is discussed later.

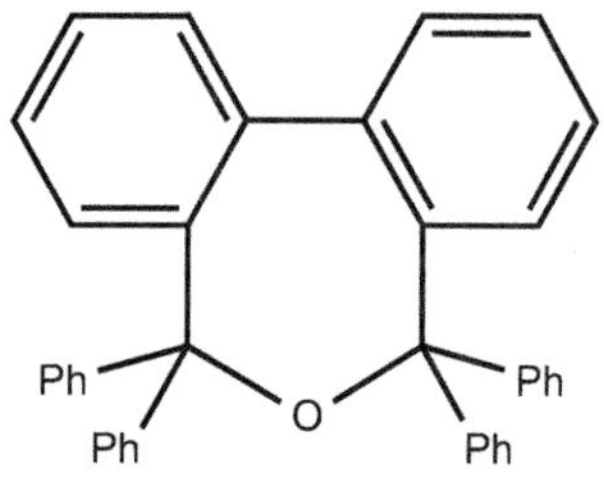

Fig. 1.17 The organic oxepin molecule used as the first example of density and symmetry calculations.

The term 'molecular mass' is generally used, though some compounds are ionic rather than molecular; 'mole mass' is more appropriate.

The subject of space groups is, in detail, beyond the scope of this book. The space group $P2_1/n$ (an alternative setting for the conventional $P2_1/c$, with a different choice of unit cell axes) is, in fact, the most common of all 230, accounting for roughly one-third of all known molecular crystal structures; its arrangement of symmetry elements provides a particularly effective packing of molecules of general shape. Fortunately, it is one of the space groups which give a unique set of systematic absences in the diffraction pattern. The screw axis (2_1), parallel to the unit cell b axis, causes reflections $0k0$ to be absent when k is odd; the glide plane (n), perpendicular to the b axis and with its glide direction along the ac face diagonal, causes reflections $h0l$ to be absent when $h + l$ is odd. The presence of the screw axis and the perpendicular n-glide plane are indicated in the space group symbol $P2_1/n$, the capital letter P indicating a primitive (not centred) unit cell.

A non-integer value of Z must always be rounded down to a suitable integer, of course; rounding it up would correspond to solvent with negative mass.

This structure has been published: Synthesis of highly hindered oxepins and an azepine from bis-trityl carbenium ions: structural characterisation by NMR and X-ray crystallography. K.A. Carey, W. Clegg, M. R. J. Elsegood, B. T. Golding, M. N. S. Hill and H. Maskill, *J. Chem. Soc., Perkin Trans. 1* 2002, 2673–2679. The CSD REFCODE is VACYII.

To convert between grams for one unit cell (or for one molecule) and atomic mass units (officially called daltons in SI; these are masses in grams for one mole), the scale factor is Avogadro's number.

$$\text{unit cell mass} = 3.939 \times 10^{-21} \times 6.023 \times 10^{23}$$
$$= 2372 \text{ daltons} \tag{1.6}$$

The mass of one formula unit is just the sum of all the atomic masses, in this case 500.6 daltons for the formula $C_{38}H_{28}O$. From the known (or assumed!) formula mass and the experimentally determined total unit cell mass, the ratio gives the number of formula units ('molecules') per unit cell, conventionally given the symbol **Z**.

$$Z = \text{unit cell mass/formula mass}$$
$$= 2372/500.6$$
$$= 4.74 \tag{1.7}$$

This must be a whole number and appropriate to the symmetry of the crystal system and space group; the space group for this compound is $P2_1/n$, for which the expected value of Z is 4.

Clearly there is something wrong here! If the experimental measurements (unit cell geometry and density) are correct, the answer to the problem must lie in the chemical formula, which has been assumed, not proved. To find the true formula mass instead of the assumed one, we must choose an appropriate integral value for Z, in this case 4, and work backwards, from what we know to what we do not.

$$\text{formula mass} = \text{unit cell mass/}Z$$
$$= 2372/4$$
$$= 593 \text{ daltons} \tag{1.8}$$

This is 92.4 greater than the previously assumed formula mass of 500.6 and the difference is, within reasonable experimental error, equal to the molecular mass for toluene (92.1). The true complete formula of the compound is, therefore, probably $C_{38}H_{28}O \cdot C_7H_8$; for every oxepin molecule in the crystal structure, there is also one molecule of toluene solvent, incorporated during the crystallization. This gives a calculated density of 1.229 g cm^{-3}.

Such **solvent of crystallization** is by no means uncommon, and is certainly not restricted to the familiar case of water of crystallization in many salts obtained from aqueous solution (**hydrates**). Solvents often encountered in crystal structures include methanol, dichloromethane, acetonitrile, and toluene.

Since we have $Z = 4$ and this is the expected value for the space group, the asymmetric unit consists of one oxepin molecule and one molecule of toluene; molecules do not lie in **special positions** on any symmetry elements, and we can deduce nothing at this stage about the molecular shape.

As a second example, an ionic compound [(18-crown-6)K][In(SCN)$_4$(py)$_2$] (where py is pyridine) can be obtained from solution in pyridine (which serves not only as a solvent, but also as a ligand to the indium atom) (Fig. 1.18). It crystallizes in the triclinic system with $a = 8.941$, $b = 9.682$, $c = 13.113$ Å, $\alpha = 87.25$,

$\beta = 72.33$, $\gamma = 89.05°$, giving $V = 1080.3\,Å^3$. The compound is air-sensitive and there are other experimental difficulties in measuring the density. In fact, crystal densities are often not measured at all. Experience shows that, for a wide range of organic compounds and metal complexes with organic ligands, the average volume required for a molecule in a crystal structure is about $18\,Å^3$ per non-hydrogen atom (hydrogen atoms are not counted). The proposed formula for this compound has 44 non-hydrogen atoms, so a volume of about $792\,Å^3$ is expected for each cation–anion pair on the basis of this **$18\,Å^3$** rule. This is rather less than the measured unit cell volume, but more than half of it, so the unit cell cannot contain more than one formula unit, and the proposed formula is incomplete (or incorrect): the cell volume is sufficient for one cation, one anion, and two or three molecules of pyridine. There are two possible triclinic space groups: $P1$, which has no symmetry other than pure translation, and $P\bar{1}$, which has inversion symmetry. Of these, the centrosymmetric space group $P\bar{1}$ is far more common, except for structures of chiral molecules. In this space group, the asymmetric unit is half the unit cell, the other half being related to it by inversion symmetry, so the expected value of Z is 2. Since the volume calculations show that this structure has $Z = 1$ (so there is only one cation and one anion in each unit cell), both the cation and the anion must themselves have symmetry, and must lie on inversion centres. Therefore, the $[(18\text{-crown-}6)K]^+$ cation and the $[In(SCN)_4(py)_2]^-$ anion are both centrosymmetric. In the case of the anion, this means that identical ligands must occur in pairs *trans* to each other, with the indium atom sitting on the inversion centre. Hence we already know (assuming the proposed chemical formula is correct apart from additional solvent molecules, and that the much more common triclinic space group applies) that the pyridine ligands are *trans* to each other, not *cis*, one of the questions to be answered by carrying out the structure determination. There are probably two pyridine molecules for each cation–anion pair, lying in **general positions** and related to each other by inversion symmetry.

The final example is rather unusual, with a very high symmetry. Crystals of $[Cr(NH_3)_6][HgCl_5]$, obtained from aqueous solution, belong to the cubic crystal system, with $a = 22.653\,Å$; the unit cell volume is $V = a^3 = 11625\,Å^3$ and the density has been measured as $2.44\,g\,cm^{-3}$. This gives the mass of the contents of one unit cell:

$$
\begin{aligned}
\text{mass} \;&=\; \text{density} \times \text{volume} \\
&=\; 2.44\,g\,cm^{-3} \times 11625\,Å^3 \\
&=\; 2.44\,g\,cm^{-3} \times 11625 \times (10^{-8})^3\,cm^3 \\
&=\; 2.837 \times 10^{-20}\,\text{g per unit cell} \qquad (1.9)
\end{aligned}
$$

$$
\begin{aligned}
\text{then unit cell mass} \;&=\; 2.837 \times 10^{-20} \times 6.023 \times 10^{23} \\
&=\; 17090\ \text{daltons} \qquad (1.10)
\end{aligned}
$$

The mass of one formula unit is 532.0 daltons, so

$$
\begin{aligned}
Z \;&=\; \text{unit cell mass/formula mass} \\
&=\; 17090/532.0 \\
&=\; 32.1 \qquad (1.11)
\end{aligned}
$$

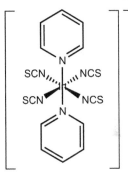

Fig. 1.18 The cation and anion of $[(18\text{-crown-}6)K][In(SCN)_4(py)_2]$.

This structure has been published: Two anionic indium(III)–thiocyanate complexes with potassium-centred complex cations. C. J. Carmalt, W. Clegg, M. R. J. Elsegood, B. O. Kneisel and N. C. Norman, *Acta Crystallogr. Sect. C* 1995, **51**, 1254–1258. The CSD REFCODE is YUXXIY.

In addition to Z, the number of chemical formula units (molecules, ion pairs, etc.) in one unit cell, crystallographers also define **Z′** as the number of chemical formula units in the asymmetric unit. The first example above had $Z′ = 1$ (the most common situation), while this second example has $Z′ = 0.5$. Structures with $Z′ > 1$ can sometimes present difficulties in their determination and interpretation, and the comparison of the chemically identical but crystallographically independent molecules is often of interest, especially when they have quite different geometries.

The space group is $Fd\bar{3}c$ (number 228 of the 230). There must, therefore, be 32 complex cations and 32 complex anions in each unit cell (there is no incorporated solvent in this example). Reference to the space group tables shows that the asymmetric unit for space group $Fd\bar{3}c$ is 1/192 of the unit cell, so a molecule or ion with no symmetry of its own would have $Z = 192$. The cation and anion here have considerable symmetry themselves, with $Z' = 1/6$. According to the tables, for $Z = 32$, the allowed point group symmetries of the ions are S_6 and D_3. Of these, the cation must have S_6 symmetry, which is consistent with essentially octahedral coordination of the Cr by six NH_3 ligands, and the anion has D_3 symmetry, which means it is a regular trigonal bipyramid with two equivalent axial and three equivalent equatorial ligands attached to Hg.

This rather extreme case shows that sometimes a great deal can be deduced about the molecular shape, even before the full structure determination is carried out (Figs 1.19 and 1.20).

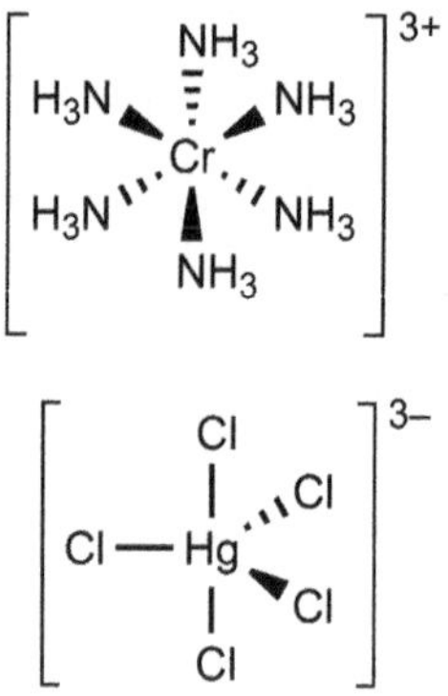

This structure has been published: Crystal structure and vibrational spectroscopy of hexaamminechromium(III) pentachloromercurate(II). W. Clegg, D. A. Greenhalgh and B. P. Straughan, *J. Chem. Soc. Dalton Trans.* 1975, 2591–2593. Having no organic content, it is not recorded in the CSD.

Fig. 1.19

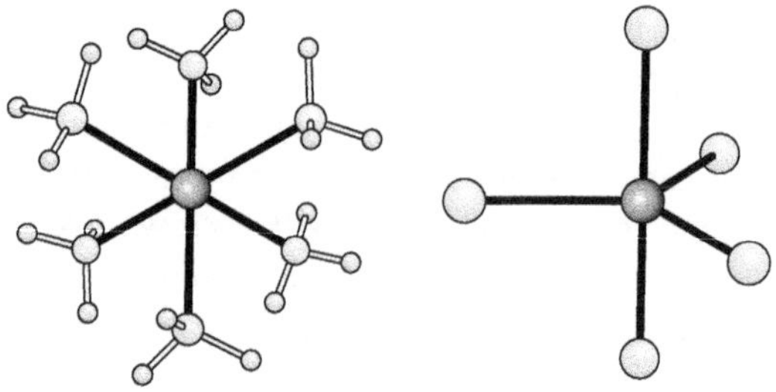

Fig. 1.20 The structures of the component ions of $[Cr(NH_3)_6][HgCl_5]$.

There are, of course, numerous instances in which the proposed chemical formula and the experimentally measured unit cell volume are just not compatible, with no possible integer value of Z, even when possible solvent of crystallization is included. In such cases, these preliminary measurements and calculations show that the material being studied is simply not what was thought. Sometimes it is starting material or a decomposition product, but sometimes it is a totally unexpected and unknown material of considerable interest. Without any other non-crystallographic information, only a full structure determination based on the measured intensities will show the answer, unless the unit cell can be recognized as that of an already known crystal structure (perhaps with the help of a computer database, as described in Chapter 2). Further examples of these calculations, in the form of problems for solution by the reader, can be found at the end of this chapter (Section 1.10).

1.7 The intensities of diffracted X-rays

Background and notation

The intensities of the diffraction pattern and the arrangement of atoms in the unit cell of the crystal structure are related to each other by Fourier transformation: the diffraction pattern is the Fourier transform of the electron density, and the electron density is itself the Fourier transform of the diffraction pattern.

The mathematical equations for crystallographic Fourier transformations have a fearsome appearance, but they can be understood in terms of the physical processes which they represent, with the help of the optical analogues presented earlier. Much of the difficulty presented by the Fourier transform equations comes from their use of **complex number** notation. This can be regarded as just a convenient way of manipulating two numbers with only one symbol. The two numerical values associated with each reflection in a crystal diffraction pattern are the *amplitude* |F| and the *phase* φ of the diffracted wave. Figure 1.21 shows two such waves; the amplitude is represented by the height of the wave, and the phase by the horizontal shift relative to some chosen origin.

Another, more compact, way of representing the same waves is shown in Fig. 1.22. Each wave is represented by an arrow with its tail at the centre of the diagram (the origin); the length of the arrow is proportional to the wave amplitude |F|, and the direction shows the phase φ, with a zero phase angle on the horizontal axis to the right and other angles (0–360° or 0–2π radians) measured anticlockwise from there. This is a vector representation: a vector **F** has both magnitude |F| and direction φ, like the arrows in the diagram.

Instead of the two values of length and direction from the origin, each of the arrowhead positions could be specified by two coordinates on the horizontal (A) and vertical (B) axes. The mathematical relationship between the vector and coordinate representations is in terms of the Pythagoras theorem and simple trigonometry (Fig. 1.23).

$$|F|^2 = A^2 + B^2; \quad \tan\phi = B/A \qquad\qquad (1.12)$$
$$A = |F|\cos\phi \; ; \; B = |F|\sin\phi$$

Forming an image of electron density from a diffraction pattern is the equivalent of the operation of a microscope lens system and involves adding together waves with their correct relative amplitudes and phases. This is shown in wave terms for just two waves in Fig. 1.24 and in vector terms in

The mathematical process of Fourier transformation is reversible, and the effect of performing Fourier transformation on a function twice in succession is to reproduce the original function, multiplied by a scale factor and by –1; this is why a simple optical microscope produces a magnified, inverted image of the object being studied.

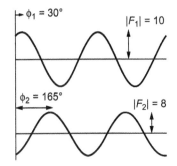

Fig. 1.21 Amplitudes and phases of two waves.

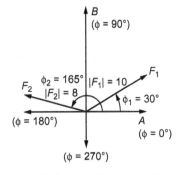

Fig. 1.22 The same two waves as in Fig. 1.21, represented as vectors.

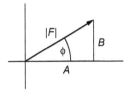

Fig. 1.23 The relationship between vector and coordinate representations.

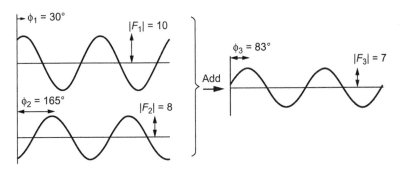

Fig. 1.24 Addition of two waves to give a resultant wave.

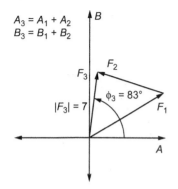

$A_3 = A_1 + A_2$
$B_3 = B_1 + B_2$

$|F_3| = 7$

$\phi_3 = 83°$

Fig. 1.25 The same wave addition as in Fig. 1.24, as a vector representation.

This means, logically, that i is the square root of -1, a difficult concept which leads to the unhelpful and misleading description (as far as our subject is concerned) of the two components as 'real' and 'imaginary': they are, in fact, both equally real!

To avoid cramped superscripts, $e^{i\phi}$ can also be written as $\exp(i\phi)$. We adopt this notation from here on.

Fig. 1.25. The A component of the resultant combined vector is simply the sum of the A components of the individual vectors, and similarly for the B components. Then the final amplitude $|F|$ and phase ϕ can be calculated from the final A and B by equations 1.12. This is true for the combination of any number of waves.

$$\text{combined } A = A_1 + A_2 + \cdots + A_n = \sum_{i=1}^{n} A_i$$

$$\text{combined } B = B_1 + B_2 + \cdots + B_n = \sum_{i=1}^{n} B_i$$

(1.13)

Clearly, the A and B components must be summed separately and not mixed up together during the process until the sums are complete. The A components are terms involving cosines of phase angles, and the B components are analogous terms involving sines of phase angles (equation 1.12).

In practice, computer programs to calculate crystallographic Fourier transforms do treat the A and B components of the individual reflections separately in this way. For convenience in showing the mathematical relationships, however, avoiding the need for two versions of every equation, the two components can be represented by a single symbol using complex number notation. A complex number has two parts, which are kept separate by multiplying one of them by the symbol i. A full treatment of complex number theory is beyond the scope or requirement of our subject. Here we need only a few of its most important features. The 'non-i-terms' and the 'i-terms' are equivalent to two orthogonal coordinates, the components of two-dimensional vectors (the horizontal and vertical axes in Figs 1.22 and 1.25). Multiplication by i is equivalent to rotating a vector by 90° anticlockwise, for example from the A axis to the B axis, so multiplying by i^2 is a 180° rotation which turns a vector **F** into its opposite vector **−F**. So we can write one symbol F for a wave, where

$$F = A + iB$$

(1.14)

From the previous relationship between the vector and coordinate representations, and using a property of complex numbers whereby $e^{i\phi} = \cos\phi + i\sin\phi$, the above equation becomes

$$\begin{aligned} F &= |F|\cos\phi + i|F|\sin\phi \\ &= |F|(\cos\phi + i\sin\phi) \end{aligned}$$

(1.15)

so $F = |F|\,e^{i\phi}$

and we have the amplitude $|F|$ and phase ϕ represented by the one symbol F, a complex number. Remember that each reflection, or diffracted wave, is labelled by its three indices hkl, so for each reflection

$$F(hkl) = |F(hkl)|\exp[i\phi(hkl)]$$

(1.16)

$F(hkl)$ is called the **structure factor** of the reflection with indices h, k, and l.

The forward Fourier transform (the diffraction experiment)

The diffraction pattern is the *Fourier transform* (FT) of the electron density. In mathematics:

$$F(hkl) = \int_{cell} \rho(xyz) \cdot \exp[2\pi i(hx + ky + lz)]dV \tag{1.17}$$

   The structure factor (amplitude and phase) for reflection *hkl* is given by taking the value of the electron density at each point in the unit cell, $\rho(xyz)$, multiplying it by the complex number $\exp[2\pi i(hx + ky + lz)]$, and adding up (integrating over the whole cell volume, $\int_{cell}dV$) these values. Positions in the unit cell are measured from one corner (the origin) and the coordinates x, y, z are in fractions of the corresponding cell edges a, b, c: for example, the very centre of the unit cell has coordinates $x = ½, y = ½, z = ½$. This calculation can be carried out mathematically to mimic the observed experimental diffraction of X-rays by a crystal. It needs to be done for each reflection and it produces a set of calculated structure factors, each with an amplitude $|F(hkl)|$ and a phase $\phi(hkl)$. In the physical experiment itself, of course, only the amplitudes are obtained.

   This equation shows how each bit of the structure contributes to every reflection. Since all the unit cells are identical, the total diffraction pattern of the crystal is just the Fourier transform of the contents of one unit cell multiplied by the number of unit cells in the crystal, so there should be just a simple scale factor between the observed and calculated sets of amplitudes.

   The equation in this form is not convenient for calculation, because it contains integration and a continuous function $\rho(xyz)$. Summation of a finite number of terms is easier. This can be achieved by expressing the electron density instead in terms of individual atoms.

   One atom scatters X-rays rather like a single circular hole scatters light passing through it (Fig. 1.14), except that the scattering is by electrons throughout the atom and not just on its edges; this means no outer rings of brightness are formed. In the forward direction ($2\theta = 0°$) all the electrons scatter X-rays exactly in phase, but at all other angles there are partial destructive interference effects, so the intensity falls off as θ increases. The variation of intensity with angle (usually shown as a function of $(\sin\theta)/\lambda$, so that it is the same for X-rays of different wavelengths) is called the **atomic scattering factor** $f(\theta)$ and has the general form shown in Fig. 1.26. It is measured in units of electrons; $f(0)$, the scattering factor for zero deflection, is equal to the atomic number. These functions are known for atoms and ions of all elements and are obtained from quantum mechanical calculations; they are available in standard reference tables and incorporated into many crystallographic computer programs.

   Atoms in crystalline solids, however, are not stationary; they vibrate, to an extent which depends on the temperature, and this effectively spreads out the atomic electron density and increases the interference effects. The atomic scattering factor falls off more rapidly with increasing angle, and is not the same for all atoms of the same element, because they generally have different amounts of vibration unless they are symmetry-equivalent. For an atom which vibrates

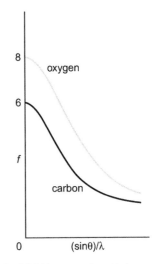

Fig. 1.26 X-ray atomic scattering factors for carbon and oxygen.

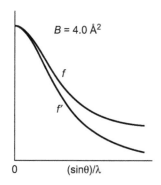

Fig. 1.27 The effect of atomic vibration on X-ray scattering factors; in this example, $B = 8\pi^2 U = 4$ Å^2.

Note that there is a diffracted wave $F(000)$, for which $\theta = 0$ and so $|F(000)|$ is the sum of all the zero-angle atomic scattering factors; this is just the total number of electrons in one unit cell. The intensity of this wave cannot be measured experimentally, because it coincides with the majority of the incident X-ray beam, which passes undeflected through the crystal. All other $|F(hkl)|$ values are smaller than $|F(000)|$, which represents all the atoms scattering together cooperatively.

The forward and reverse Fourier transforms differ in that one has a negative sign inside the exponential term. The unmeasured term $F(000)$ must also be included; it is equal to the total number of electrons in one unit cell.

equally in all directions (**isotropic** vibration), the effect is to multiply the atomic scattering factor by a term containing an **isotropic displacement parameter** U (see Fig. 1.27), which represents a mean-square amplitude of vibration for the atom, a measure of how much it is vibrating.

$$f'(\theta) = f(\theta) \cdot \exp\left(\frac{-8\pi^2 U \sin^2 \theta}{\lambda^2}\right) \tag{1.18}$$

Note that U has units Å^2 and the extra term has a value < 1. The larger the value of U, the more the curve falls off at higher Bragg angles.

With discrete atoms instead of a continuous electron density function, the forward Fourier transform takes the form

$$F(hkl) = \sum f_j(\theta) \cdot \exp(-8\pi^2 U_j \sin^2 \theta / \lambda^2) \cdot \exp[2\pi i(hx_j + ky_j + lz_j)] \tag{1.19}$$

$$\mathbf{1} \qquad \mathbf{2} \quad \mathbf{3} \qquad\qquad \mathbf{4} \qquad\qquad\qquad \mathbf{5}$$

The summation is made over all the atoms in the unit cell, each of which has its appropriate atomic scattering factor f_j (a function of the Bragg angle θ), a displacement parameter U_j, and coordinates (x_j, y_j, z_j) relative to the unit cell origin. This summation must be carried out for every diffracted wave $F(hkl)$.

Although equation 1.19 looks complicated, it can be readily understood in terms of the physical process it represents. Every atom scatters X-rays falling on it (terms **3** and **4** in the equation). In any particular direction (hkl), these separate scattered waves from each atom have different relative phases which depend on the relative positions of the atoms (term **5**), and the total diffracted wave in that direction (term **1**) is just the resultant sum (term **2**) of the X-rays scattered by the individual atoms. The equation just represents the combination or addition of many waves to give one resultant wave in each direction; the graphical equivalent for two waves was given in Fig. 1.24.

The reverse Fourier transform (the recombination calculation)

The electron density is the *reverse Fourier transform* (**FT⁻¹**) of the diffraction pattern. Because the diffraction pattern of a crystal consists of discrete reflections rather than a diffuse pattern, this Fourier transform is a summation, not an integral.

$$\rho(xyz) = \frac{1}{V} \sum_{h,k,l} F(hkl) \cdot \exp[-2\pi i(hx + ky + lz)]$$

$$or\ \rho(xyz) = \frac{1}{V} \sum_{h,k,l} |F(hkl)| \cdot \exp[i\phi(hkl)] \cdot \exp[-2\pi i(hx + ky + lz)] \tag{1.20}$$

$$\mathbf{1} \qquad\quad \mathbf{2} \qquad \mathbf{3} \qquad\qquad \mathbf{4} \qquad\qquad \mathbf{5}$$

Remember that $F(hkl)$ is a complex number, containing both amplitude and phase information, as is shown explicitly in the second version. The term $1/V$ is necessary in order to give the correct units (structure factors, like atomic scattering factors, have units of electrons, but electron density is electrons per Å^3).

The summation is performed over all values of h, k, and l, i.e. all the reflections in the diffraction pattern contribute to it. In practice, reflections are measured only to a certain maximum Bragg angle, but this is usually unimportant because the higher angle reflections are weaker and so contribute relatively little to the sums. The summation must be carried out for many different coordinates x, y, z in order to show the variation of electron density in the unit cell and hence locate the atoms where the electron density is concentrated in peaks.

As for the forward Fourier transform, this equation is readily understood in terms of the (unachievable!) physical process it represents. The image of the electron density (**1**), which originally generated the diffraction pattern, is obtained by adding together (**2**) all the diffracted beams, with their correct amplitudes (**3**) and phases (**4**, **5**); the correct relative phases here include the intrinsic phases of the waves themselves, relative to the original incident beam (**4**), and an extra phase shift appropriate to each geometrical position in the image relative to the unit cell origin (**5**).

The relative phase shifts (**5**) can be calculated as required, but the intrinsic phases $\phi(hkl)$ of the different reflections are unknown from the diffraction experiment. This means that it is not possible simply to calculate the reverse Fourier transform once the diffraction pattern has been measured. Here, once again, in the mathematical basis of the method we see the nature of the 'phase problem'.

1.8 Sources of X-rays

So far, uses of X-rays have been discussed, but nothing about how they are produced. Details are not important for our purposes, and only a brief outline is given here for completeness.

In most laboratories the standard source of X-rays is an **X-ray tube** (Fig. 1.28). Until fairly recently this was usually an evacuated enclosure of glass (or ceramic material) and metal construction which produces electrons by passing an electrical current through a wire filament, accelerates them to a high velocity by an electrical potential of typically 40 000–60 000 volts across a few millimetres, then stops them dead with a water-cooled metal block. Most of the electron kinetic energy is converted to heat and wasted, but a small proportion generates X-rays by interaction with the target metal atoms. Among other effects occurring, if an electron in a core atomic orbital is ejected (ionized), an electron from a higher orbital can take its place and the drop in energy produces emission of radiation of a definite frequency and wavelength ($\Delta E = h\nu = hc/\lambda$). Several such transitions are possible, so the output of radiation from the target consists of a series of intense sharp maxima, superimposed on a broad-spectrum background of radiation from non-quantum processes (Fig. 1.29). One particular peak, usually the most intense, can be selected and the rest of the output suppressed by exploiting the Bragg equation: the beam of radiation falls on a single crystal of known structure (often graphite) suitably oriented so that the desired wavelength satisfies the equation; only this wavelength is diffracted at the appropriate angle, the others pass straight through the **monochromator** crystal. The most commonly used X-ray tube target materials are copper and molybdenum, which give characteristic X-rays of wavelengths 1.54184 and 0.71073 Å respectively.

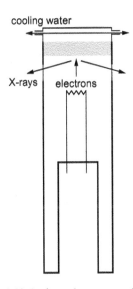

Fig. 1.28 A schematic representation of a conventional laboratory X-ray tube.

Here, h is Planck's constant, not to be confused with hkl indices of X-ray reflections!

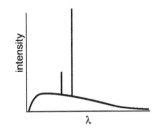

Fig. 1.29 The spectrum of X-rays emitted by an X-ray tube.

Various developments of the basic X-ray tube produce higher intensities. The main limitation is the amount of heat produced. For a more powerful electron beam, the target must be constantly moved in its own plane to spread the heat load, producing 'rotating anode' X-ray tubes, which can provide about one order of magnitude more intensity. Alternatively, unwanted melting of the target can be avoided by using a liquid metal target in the form of a jet of an alloy of gallium, indium, and tin, which is a liquid at room temperature; the wavelengths of X-rays produced from gallium and indium are 1.3414 and 0.5151 Å respectively.

An alternative approach is to focus the electron beam in the X-ray tube with electric and/or magnetic fields so that the X-rays are generated from a much smaller spot on the target; in such microfocus tubes, a higher X-ray intensity can be obtained from a much lower electron current, reducing both the power consumed and the heat generated. The X-rays can be collected, monochromated, and effectively focused by a range of available advanced X-ray optics components including curved and graded multilayer materials which have a gradual change of chemical composition, and hence of lattice spacing, in one dimension, and planar or curved grazing-incidence mirrors. Combinations of these recent technological developments can provide increases in X-ray intensity of several orders of magnitude over conventional X-ray tubes.

Even more intense X-rays, as well as other parts of the electromagnetic spectrum, are produced in a *synchrotron storage ring* (Fig. 1.30), in which electrons (or positrons) moving at almost the speed of light are constrained by magnetic fields to move in a circle usually hundreds of metres in diameter. The radiation, emitted tangentially from the ring, has a continuous spectrum, ranging from infrared to X-rays, from which a single wavelength of any value can be selected by a monochromator or other optics (the X-rays are 'tuneable') and is many orders of magnitude more intense than the output of laboratory X-ray sources. Such facilities, of course, are vastly more expensive and are national or international facilities with a wide variety of other scientific applications in addition to X-ray diffraction.

While the bending magnets of a **synchrotron** source themselves produce X-rays, and these have been successfully exploited in so-called second-generation synchrotrons from around 1980 onwards, most synchrotron sources in current operation are third-generation, in which the main X-ray output is from complex arrays of many magnets (known as wigglers and undulators depending on their detailed construction) located in the straight sections between bending magnets and hence called **insertion devices**. A fourth generation of sources is now being planned and constructed, based on very long accelerators and undulators and known as X-ray free-electron lasers (XFELs); these will give very short and rapid pulses of extremely high-intensity X-rays and will further extend the power of crystallography and other sciences.

Although synchrotron sources have many special properties that can be exploited in a wide range of experiments related to crystallography (and also in other forms of scattering, spectroscopy, imaging and other applications), the most important points for our purpose here are that they provide extremely high intensities of X-rays, these X-ray beams are usually finely focused for beneficial

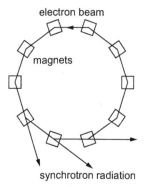

electron beam

magnets

synchrotron radiation

Fig. 1.30 The production of synchrotron radiation from relativistic electrons in a special type of particle accelerator.

First-generation synchrotron sources were built primarily as particle accelerators and colliders for use in high-energy physics, and their production of electromagnetic radiation was considered an unfortunate loss of energy by their main users; in some cases it could be used without significant disturbance of the main operational purpose. Second-generation sources were designed with the stable and reliable production of synchrotron radiation as their main aim, and would be better referred to as storage rings; the radiation is produced by the bending magnets. Third-generation sources make extensive use of insertion devices, though the bending magnets also provide useful output.

use with very small crystals, and the X-ray wavelength can be selected through optical components from the broad spectrum available. Some of the benefits will be demonstrated later with case studies (Chapter 3) and with reference to particular problems and applications.

1.9 Summary

- X-ray crystallography is based on the diffraction of X-rays by a crystalline material, a cooperative form of scattering, in contrast to spectroscopic methods, most of which are based on the absorption (or emission) of electromagnetic radiation.

- X-rays are used because they have a wavelength comparable to the size of atoms and molecules, giving rise to measurable diffraction effects from crystals. The process resembles the operation of a microscope using visible light, but the recombination of scattered X-rays by a conventional lens system is not possible, so it has to be done mathematically. Unfortunately, only the diffracted X-ray amplitudes (as intensities) are available from the recorded diffraction pattern, while the relative phases are lost.

- The fundamental property of the crystalline state is translation symmetry, characterized by the concepts of the lattice and unit cell. Crystalline solids cannot display some symmetry elements possible for single molecules, but they can show other symmetry elements, with translational components, that do not occur in finite molecules. Solid-state symmetry is described by space groups, of which there are 230.

- A diffraction pattern from a single crystal consists of discrete 'reflections' in a regular pattern. The geometry of a diffraction pattern is related to the lattice and unit cell geometry through the Bragg equation. The symmetry of a diffraction pattern is related to the space group symmetry of the crystal structure. The intensities of a diffraction pattern arise from the detailed contents of the unit cell, *i.e.* the identity and positions of atoms in the structure, the relationship between these being Fourier transformation: the diffraction pattern is the Fourier transform of the crystal structure, and *vice versa*.

- X-rays are generated from the impact of fast-moving electrons on a metal target in a range of laboratory equipment; the most intense X-rays are available from national and international synchrotron and free electron laser facilities.

1.10 Exercises

Exercise 1.1

Why are X-rays used, rather than any other part of the electromagnetic spectrum, for crystal structure determination? Why is it not possible to build and use an X-ray microscope to observe molecules directly? Why is a single crystal used for the experiment?

Exercise 1.2

For the cubic crystal system, many calculations are easier than in lower-symmetry systems. For example, the spacing of lattice planes d_{hkl} is simply $a / \sqrt{(h^2 + k^2 + l^2)}$. For a cubic unit cell with $a = 10\,\text{Å}$, calculate the d spacings for the lattice planes (1 0 0), (2 0 0), (0 2 0) and (1 1 1). Using the Bragg equation (1.3), calculate the Bragg angle θ for the reflections from these lattice planes, with an X-ray wavelength $\lambda = 0.7\text{Å}$.

Exercise 1.3

What is the smallest observable d spacing in a diffraction pattern measured with X-rays from a copper target ($\lambda = 1.54184\,\text{Å}$)? What implication does this have for the feasibility of resolving individual atoms in an electron density map?

Exercise 1.4

The complex $[(C_{18}H_{18}N_4S)HgBr_2]$ (relative molecular mass 682.8) crystallizes from solution in acetonitrile (CH_3CN) to give triclinic crystals with a unit cell volume of $1113.5\,\text{Å}^3$ and with a density of $2.16\ \text{g cm}^{-3}$. Calculate the number of molecules of complex per unit cell and the number of molecules of solvent per unit cell.

Exercise 1.5

A compound of supposed formula $K^+[In(NCS)_4(bipy)]^-$, where bipy is the chelating ligand 2,2'-bipyridyl, is obtained from solution in THF (C_4H_8O) as monoclinic crystals with $a = 14.985\,\text{Å}$, $b = 17.375\,\text{Å}$, $c = 16.437\text{Å}$, $\beta = 92.23°$. The density of the crystals is $1.40\ \text{g cm}^{-3}$, and the relative molecular mass for the above formula is 542.4. Calculate the unit cell volume and deduce the number of cations and anions per unit cell (expected values are 2, 4, or 8) and the number of molecules of THF per cation (which are likely to be coordinated to it).

X-ray crystallography in practice

2.1 Introduction

Having examined the physical basis of X-ray crystallography and its expression in mathematical notation, and thereby outlined the main principles and concepts of the subject, we consider in this chapter how the method works in practice. The various successive steps of a typical crystal structure determination are described in general terms and illustrated with appropriate examples. A number of common potential problems are described together with ways of dealing with them.

Fig. 2.1 shows an outline of crystal structure determination in a simplified form as a schematic flowchart. The steps involved are in the boxes. To the right of each is listed the information obtained and to the left an indication of the timescale involved in carrying out the operation. Some of these times vary considerably, depending on the quality of the sample being studied, the resources available for the work, the size and complexity of the structure, the skill of the crystallographer, and a certain amount of luck, and most of them become shorter as techniques advance.

2.2 The preparation and selection of samples

The sample must be a **single crystal**, in which all the unit cells are identical and are aligned in the same orientation, so that they scatter cooperatively to give a clear diffraction pattern consisting of individual X-ray beams, each in a definite direction. Outward appearance such as regularity of shape is not important, but rather the internal regularity of the molecular arrangement on a well-defined lattice; many single crystals have an unpromising irregular shape, while polycrystalline and even non-crystalline materials such as glass may have beautiful external forms.

The intensities of X-rays diffracted by a crystal depend on the crystal size, the unit cell volume, and the types and numbers of atoms in the unit cell, as well as on the incident beam intensity; an approximate formula for the relative scattering power of a crystal is as follows.

$$\frac{V_{crystal}}{V_{cell}^2} \sum_{cell} f^2 \tag{2.1}$$

In reality, all crystals have faults in their internal structure, so unit cells are not exactly aligned. The range of misalignment is called the **mosaic spread**, because the slightly misaligned sub-microscopic blocks of a real single crystal resemble the tiles in a mosaic (Fig. 2.2). For a good quality single crystal, the mosaic spread is only a fraction of a degree.

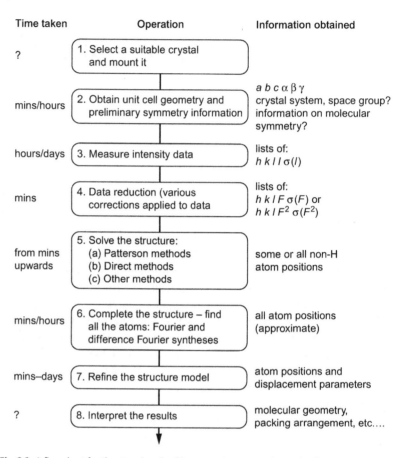

Fig. 2.1 A flowchart for the steps involved in a crystal structure determination.

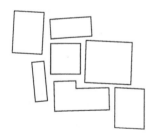

Fig. 2.2 Mosaic structure of a single crystal (highly exaggerated).

The diffracted intensities are directly proportional to the crystal volume, but X-rays are also absorbed by crystals and this effect increases exponentially with crystal dimensions; **X-ray absorption** affects the measured intensities, introducing a systematic error, for which a correction may need to be made (see Section 2.6). The amount of absorption (as well as the intensities of diffraction) depends on the X-ray wavelength and on the chemical composition, and can be very high when heavier elements are present. Systematic errors are also produced if the crystal is not completely bathed in the incident X-ray beam throughout the diffraction measurements, and most X-ray beams are less than 1 mm in cross-section; those from microfocus tubes and synchrotron sources can be 100–200 μm (microns) or even smaller. A typical acceptable crystal size for a conventional X-ray source is a few tenths of a millimetre; a smaller size and uniform dimensions are preferable for samples containing heavy atoms, and very small crystals can be examined with intense synchrotron radiation. Such crystals, much smaller than the popular image, usually need to be examined and handled under a microscope. A microscope with polarizing filters provides some useful optical tests of the quality of a crystal, but the ultimate test is the X-ray diffraction pattern. Crystals can be cut with a sharp scalpel, but this sometimes adversely affects the crystal quality.

Suitable crystals are sometimes produced in the initial synthesis of a compound, but often recrystallization is necessary. This process can be difficult, unpredictable, frustrating, and time-consuming and is not guaranteed to succeed; it is often described as an art rather than a science. The objective is quite different from that of recrystallization in synthesis—good quality single crystals of suitable size, with high yield not a priority—though both aim for a pure material, and special techniques have been devised. It is not uncommon for a crystal structure to incorporate molecules of solvent, so the solvent itself is one of the conditions which can be varied in the quest for suitable crystals.

One single crystal is separated from the rest of the sample and is mounted on a device which will hold it firmly in the X-ray beam; a precision of hundredths of a degree is required. Since the diffraction experiment involves rotating the crystal in the beam during exposure, as explained in Section 2.3, lateral adjustments need to be available to position the crystal accurately on each rotation axis. For some less commonly used equipment, it is an advantage if one unit cell axis can be aligned in a particular direction, so there may also be provision for angular adjustments. Such a device, known as a **goniometer head**, is shown in Fig. 2.3. Apart from the sample itself, no crystalline material should be in the X-ray beam, so the crystal is usually attached (with a minimum quantity of an amorphous adhesive) to a fine glass fibre attached to the goniometer head (Fig. 2.4), or to a specially designed thin plastic mount or a fine fibre loop. The glue and mount contribute a small amount to general background scattering but not to the sharp diffraction maxima.

Samples which are air-sensitive or which degrade by loss of loosely bound solvent require special treatment. Handling them in an inert-atmosphere glove-box is possible but difficult, especially if the crystals are very small. They may be sealed in thin-walled glass capillary tubes, an operation which is considerably easier if brief exposure to the air can be tolerated. Alternatively, the crystals can be coated with an inert viscous oil and then manipulated without difficulty under a normal microscope in the open atmosphere; if the X-ray examination is to be carried out at a sufficiently low temperature that the oil vitrifies to a glass, it can be used simultaneously as an adhesive and a protective coating, and this provides a particularly elegant and simple solution for materials of even extreme air sensitivity (Fig. 2.4). Since collecting diffraction data at low temperature usually leads to a better result because of reduced atomic motion in the crystal (and is now considered the norm rather than the exception), this technique is widely used, whether samples are air-sensitive or not.

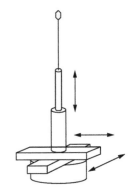

Fig. 2.3 Schematic representation of a goniometer head with a mounted crystal. Such devices were originally used on equipment for the optical measurement of angles between well-developed flat faces of crystals, which is the derivation of the word goniometer (angle-measuring device).

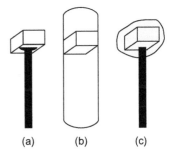

(a)   (b)   (c)

Fig. 2.4 Crystal mounting methods: (a) glued to a fine glass fibre (or other non-crystalline mount); (b) enclosed in a capillary tube; (c) coated with an inert oil for subsequent cooling.

2.3 Measuring diffraction patterns

An X-ray diffraction experiment involves several equipment components: a source of X-rays, a suitable mounted crystal, a device for rotating the crystal around one or more axes in the X-ray beam (variously called a camera, goniometer, or diffractometer), an X-ray-sensitive detector, possibly a device for cooling (or heating) the crystal during the experiment, and computer control for the

It is also possible, in more specialized experiments, to control the pressure and/or the chemical environment of the crystal, such as providing a particular gaseous atmosphere.

various movements and for storage of the measured data. In order to interpret and use a diffraction pattern, indices *hkl* must be assigned to each individual reflection; to achieve this, the unit cell must be obtained, together with a knowledge of its orientation relative to the goniometer head on which the crystal is mounted. Three different types of equipment widely used at different stages in the development of X-ray crystallography, together with examples of their measured diffraction patterns, are shown in Fig. 2.5.

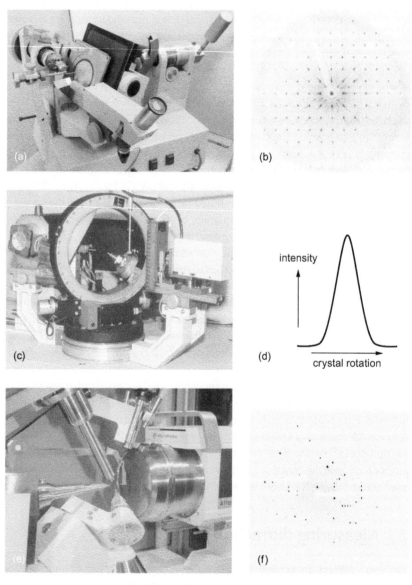

Fig. 2.5 (a) One type of X-ray camera and (b) an example of a photograph produced by it; (c) a serial four-circle diffractometer and (d) a typical single reflection profile obtained by rotating the crystal through the correct Bragg setting; (e) a modern area-detector diffractometer and (f) an image recorded for a few seconds with a small crystal movement.

For about the first 50 years of X-ray crystallography, diffraction patterns were usually recorded on photographic film (Fig. 2.5(a) and (b)). Although this is now mainly of historical interest, there are some basic features of film-based measurements that largely disappeared when so-called serial diffractometers became commonplace from around the 1960s but returned in modified form with the widespread introduction of area-detector diffractometers in the 1990s.

The diffraction conditions represented by the Bragg equation are severe, and will be satisfied for only very few reflections for a randomly oriented stationary crystal in an X-ray beam, because few of the lattice planes will fortuitously be oriented at the correct θ angle, so the pattern recorded on film (or an area detector) will show only a few spots (Fig. 2.6(a)). In order to bring more lattice planes into a reflecting position, the crystal must be rotated in the X-ray beam (Fig. 2.6(b)). Recording the whole of the diffraction pattern on one film, however, leads to severe overlap of the reflections occurring at different stages of the rotation, because three-dimensional information is being compressed into a two-dimensional record, and its measurement and interpretation are impossible (Fig. 2.6(c)).

Instead, selected portions of the diffraction pattern need to be recorded separately on different films, or at different times on an area detector. The interpretation of photographically recorded patterns is greatly assisted if the rotation of the crystal is about the direction of a unit cell axis, and further simplification results from some types of correlated movement of the film with that of the crystal, and from the use of metal screens of appropriate shape that intercept most of the reflections and allow through to the film only those belonging to one particular two-dimensional slice through the three-dimensional diffraction pattern. Assigning *hkl* indices to the individual reflections is then a simple matter of counting along obvious rows of spots. Several types of X-ray camera have been developed over many years to achieve such effects, each operating with a particular geometrical combination of crystal orientation, film and screen shape, and film movement. The geometry, symmetry, and intensities of the diffraction pattern can all be obtained from a suitable set of photographs, and many earlier crystal structures were determined in this way, but electronic X-ray detectors offer major advantages of speed, precision, and convenience.

An **X-ray camera** is an instrument for recording X-ray diffraction patterns on photographic film. A **diffractometer** is an instrument which records diffraction patterns by means of some kind of X-ray-sensitive detector other than photographic film, usually involving the conversion of incident X-ray energy into an electronic signal, possibly via visible light in a two-stage process.

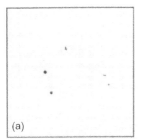 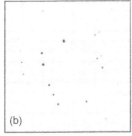 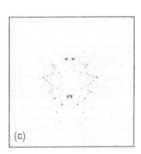

(a)   (b)   (c)

Fig. 2.6 (a) A diffraction pattern recorded on an electronic area detector from a stationary crystal; a similar pattern would be obtained on a photographic film, which is effectively a non-electronic area detector; (b) a diffraction pattern recorded from a 5° rotation of the same crystal; (c) a diffraction pattern recorded from a full 360° rotation of the same crystal.

A scintillation counter contains a material, such as thallium-doped sodium iodide, which produces light in the visible region when X-rays fall on it. The light is detected and the signal amplified by a photomultiplier, so that the overall effect is an electrical pulse for each X-ray photon incident on the face of the detector.

From around the 1960s, computer-controlled diffractometers became the standard means of collecting diffraction data. Instead of photographic film, an electronic device is used which is sensitive to X-rays. The most commonly used detector for about the next 30 years, a *scintillation counter*, is typically a few millimetres in diameter and so is capable of detecting and measuring only one reflection at a time; such devices are known as serial diffractometers. For each reflection, the detector must be moved round one axis (usually vertical) to the correct 2θ angle. Because the detector can see only reflections which occur in the horizontal plane, more than one axis of rotation is needed for the crystal. The most widely used types of diffractometer have three rotation axes for the crystal, giving more than enough freedom, so that there is even a choice of settings possible for many of the reflections. With one of these *four-circle diffractometers* reflections (positions and intensities) are observed one at a time, the crystal and detector being moved under computer control from each one to the next in sequence (Fig. 2.5 (c) and (d)).

More recently, X-ray detectors became available which record over a considerably larger area and are position-sensitive: a number of incident beams can be recorded at the same time, and their positions as well as intensities are known. There are various types of **area detectors** based on different technologies, each with particular advantages and disadvantages of size, sensitivity, spatial resolution, speed of read-out, and cost; they can be regarded simply as electronic equivalents of photographic film in many respects. An area detector can replace the scintillation counter of a four-circle diffractometer, but it is also possible to reduce the number or range of rotation axes for the crystal, because of the size of the detector; it is no longer necessary to bring all reflections into the horizontal plane in order to record them (Fig. 2.5 (e) and (f)).

Two particular types of area detectors are currently in widespread use. The majority of diffractometers use a charge-coupled device (CCD), similar to the sensor in a digital camera or camera phone; the CCD does not directly detect and record X-rays, but relies on an intermediate phosphor sensor that converts the X-ray energy to light. A CCD is an integrating device, in the sense that the incident signal is built up over a period of time (usually seconds) and then read out electronically, so recording and read-out alternate in its operation and the crystal needs to be moved in steps to record a complete diffraction pattern. More recent so-called pixel detectors record the incident X-rays directly and instantaneously, with simultaneous read-out, so they can operate continuously while the crystal is rotated at a constant speed; they are much faster and are capable of recording a wider range of intensities (a higher dynamic range) without overloading their capacity. They are particularly valuable for synchrotron sources. They will undoubtedly displace CCD detectors in most crystallography applications in coming years.

2.4 Obtaining unit cell geometry and symmetry

Both photographically and with a diffractometer the unit cell geometry can be measured from a preliminary subset of the complete diffraction pattern. The key step is assigning the correct indices *hkl* to each of the observed reflections. From these and the measured Bragg angle for a few reflections, the six unit cell parameters can be calculated via the Bragg equation and modified versions of it appropriate to the geometry of the particular camera or diffractometer being used.

With diffractometers, the crystal is usually mounted in a random orientation, and this has to be determined as well as the unit cell geometry. With a serial diffractometer, some tens of reflections of moderate to high intensity are located by simply driving the various motors while monitoring the detector output for a signal significantly above background (a blind search, all under computer control); with an area detector, a small number of initial images usually provides a few hundred suitable reflections. From their positions, the crystal orientation, unit cell geometry, and reflection indices have to be determined simultaneously, by calculations which are not simple and are usually regarded as computer 'black-box' methods, but they are all based essentially on the Bragg equation. The process usually takes only a few minutes. With a serial diffractometer, it is necessary to have the unit cell and orientation before the complete set of reflection intensities can be measured, but an area-detector machine can be set to collect all available data (assuming no symmetry) without this knowledge and the unit cell can be derived from the full set of data afterwards.

At this stage, it may be possible to assign the correct space group by comparison of intensities which are equivalent by symmetry, and by noting that certain special subsets of reflections have zero intensity (Fig. 2.7), which is an effect of symmetry elements with a translation component (glide planes and screw axes), but the decision is made more reliably on the basis of the complete data set later. We have already seen examples of how this may provide some information about the structure, such as molecular symmetry or presence of solvent.

Of course, the initial examination of a crystal with X-rays also shows the quality of the diffraction pattern, from which a decision is made whether to proceed with the full experiment (and how to do this) or look for a better crystal.

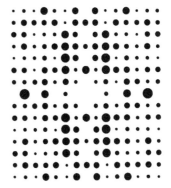

Fig. 2.7 Part of a diffraction pattern (the central section of Fig. 1.13) showing systematic absences (alternate reflections missing on the central horizontal and vertical rows) due to screw axes.

2.5 The measurement of intensities

Although diffraction intensities can be measured from photographic films, this is now rarely done. It involves estimating the degree of blackening in each spot, which can be achieved either by visual comparison with a calibrated scale or by measuring the absorption of a beam of light passed through the film. Some reflections are too weak to be seen above the general level of background scattering on the film, and these are labelled as 'unobserved'; usually no numerical value is recorded for their intensities, and they are not used in the successive calculations. The process of estimating intensities may take several weeks, depending on the size and symmetry of the structure (and hence the number of photographs required to record the complete data set) and the overall intensity of diffraction.

A four-circle diffractometer measures intensities one at a time in an automatic, computer-controlled serial process. For each reflection the crystal and detector are driven to the appropriate positions to satisfy the Bragg equation and bring the diffracted beam into the detector in the horizontal plane, and the total 'integrated intensity' is measured while the crystal is rotated through a small angle from one side of the Bragg position to the other to allow for the mosaic spread of the crystal, which produces a peak profile of a few tenths of a degree rather than a sharp spike of intensity at one angle (Fig. 2.5(d)). Some diffractometer systems carry out a detailed statistical analysis of the reflection profile shape, which provides more reliable results for weaker reflections.

With an area-detector diffractometer, many diffracted beams are recorded simultaneously. Usually the crystal is rotated about one axis, and several such scans are performed in order to obtain the complete diffraction pattern. With a CCD detector, each exposure covers a small angular range; reflections are usually spread over more than one image and sophisticated computer analysis of large quantities of data is required. With a pixel detector, continuous scanning and simultaneous read-out are possible, giving much faster data collection. Intensities of reflections are extracted from the raw images by sophisticated profile-fitting, integration and background subtraction techniques. Among other advantages, area detectors usually provide a high degree of redundancy of symmetry-equivalent data and of the same reflections measured more than once in different crystal orientations.

The crystallographer must make some decisions about the data collection procedure. These include the maximum Bragg angle to be measured (reflections at higher angles are generally weaker but add to the precision of the final structure if they have measurable intensities), the time to spend on each measurement (each single reflection, image, or scan), and whether to ensure only the complete coverage of the **unique set** of data (a fraction of the total pattern depending on the space-group symmetry) or to include more symmetry-equivalent reflections, which takes longer but again improves the quality of data overall and gives a confirmation of the symmetry. The time taken to collect the intensity data on a four-circle diffractometer depends very much on these decisions and on the size of the structure; a larger structure gives more reflections to the same maximum Bragg angle. It takes at least overnight and usually several days. Data collection typically takes only a few hours with a CCD, minutes with a pixel detector, independent of the size of the structure, since a larger structure just gives more simultaneous reflections, but longer exposures are advisable for weakly scattering samples. Here the use of high-intensity X-ray sources and particularly synchrotron radiation brings a substantial improvement.

The result of this process, from whatever equipment is used, is a list of reflections, usually thousands of them, each with *hkl* indices and a measured intensity. In addition, from diffractometer measurements, each intensity *I* has an associated **standard uncertainty** (s.u.), $\sigma(I)$, which is calculated from the known statistical properties of the X-ray generation and diffraction processes, and is a measure of the precision or reliability of the measurement. Other information available includes the directions of the incident and diffracted beams for each reflection, together with details of its position on the detector face, time of recording, etc., for use in calculating corrections for absorption and other effects.

The **unique set** of data, up to a particular maximum Bragg angle, is the total set of reflections which are independent of each other by symmetry. Application of symmetry to the unique set produces all the reflections which can be measured. For example, in the centrosymmetric triclinic case the diffraction pattern has only inversion symmetry, and the unique set is exactly half of the total available data: each reflection *h*, *k*, *l* is equivalent to the reflection *−h*, *−k*, *−l*. For a centrosymmetric monoclinic structure, the unique set is one quarter of the total, and for centrosymmetric orthorhombic it is one eighth.

A previous term for standard uncertainty, still in wide use, is *estimated standard deviation* (e.s.d.). It is an estimate of the spread of values which would be obtained if the measurements were repeated many times.

2.6 Data reduction

We have previously seen that the intensity of an X-ray beam is proportional to the square of the wave amplitude. The measured intensity is affected by various factors, however, for which corrections must be applied. The conversion of intensities I to 'observed structure amplitudes' $|F_o|$ (o = observed) or F_o^2 and, correspondingly, of s.u.s $\sigma(I)$ to $\sigma(F_o)$ or $\sigma(F_o^2)$ is known as **data reduction** and has several components.

$$I(hkl) \propto |F(hkl)|^2$$

There are corrections associated with the data collection process, which are geometrical in nature. These are a function of the geometry of the equipment and so are instrument-dependent. There is also a correction needed because reflected radiation is partially polarized (a phenomenon exploited in the use of polaroid sunglasses, for example). These geometrical corrections, known as *Lorenz-polarization* factors, are well known, and are easily and routinely made.

A correction may also be needed for changes in the incident X-ray beam intensity or in the scattering power of the crystal during the experiment. The former is particularly important for synchrotron radiation, which may fluctuate somewhat or decay gradually, depending on the details of the synchrotron operation, and the latter may be caused by some decomposition or physical deterioration of the sample in the high-energy X-ray beam. The effect of both is to make intensities change with time or with image sequence number. A correction can be made for serial diffractometer data on the basis of reflections which are measured repeatedly at intervals during the data collection to monitor changes. For area detectors, these corrections are usually made at the same time as those for absorption and related effects.

Where absorption effects are significant an appropriate correction must be made. Each reflection is affected differently by absorption, because the absorption depends on the path length of the X-rays through the crystal, and this varies as the crystal orientation is changed (Fig. 2.8). Many different types of absorption correction are used. Some are based on careful measurement of the crystal shape and dimensions and calculation of path lengths; others are based on comparison of intensities of symmetry-equivalent reflections, which should be equal but are not because of absorption effects. The high degree of redundancy of equivalent and repeated reflections from area detectors provides a convenient basis for the simultaneous correction of effects due to incident intensity variations, crystal absorption of X-rays, non-uniform exposure of the entire crystal to the X-rays, and other factors. Individual scale factors for the diffraction images are refined together with other parameters that model absorption effects (spherical harmonics, familiar to chemists as the mathematical functions for atomic orbitals, are often used) to make the intensities of symmetry-equivalent reflections as nearly equal as possible.

The data reduction process may also include the merging and averaging of repeated and symmetry-equivalent measurements in order to produce a unique, corrected, and scaled set of data (though sometimes this is performed by programs used later in solving and refining the structure). This calculation affords a numerical measure of the agreement among equivalent reflections, which is

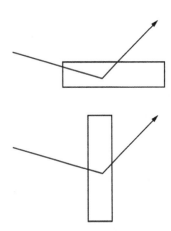

Fig. 2.8 The effect of absorption for a needle-shaped crystal.

one indication of the quality of the data and the appropriateness of the applied corrections.

At the same time, statistical analysis of the complete unique data set can provide an indication of the presence or absence of some symmetry elements, particularly whether the structure is centrosymmetric or not, though this is not infallible; and the observed overall decay of intensity with increasing $(\sin \theta)/\lambda$ gives an average atomic displacement parameter.

The various corrections for the intensities are applied also to their s.u.s. The result of this whole process, which usually takes only a matter of minutes on a computer, is a list of reflections as $h, k, l, |F_o|, \sigma (F_o)$ [or $h, k, l, F_o^2, \sigma(F_o^2)$]; one advantage of retaining the squared form is that no special treatment is required for intensities measured as negative.

In reality, of course, intensities cannot be negative, but the weakest reflections from a diffractometer may be insignificantly above background and, through statistical variations in the measuring process, may be recorded as below background, i.e. apparently net negative. These would be 'unobserved' by photographic methods. The fact that they are weak is actually valuable information in structure determination.

2.7 Solving the structure

Having measured and appropriately corrected the diffraction data, we turn now to the solution of the structure, in which we obtain atomic positions in the unit cell from the data. Remember that the objective here is to imitate a microscope lens system, recombining the individual diffracted beams to give a picture of the electron density distribution in the unit cell.

$$\rho(xyz) = \frac{1}{V}\sum_{h,k,l} |F(hkl)| \cdot \exp[i\phi(hkl)] \cdot \exp[-2\pi i(hx + ky + lz)] \qquad (1.20)$$

The mathematical expression of this process is equation 1.20, repeated above. The amplitudes $|F(hkl)|$ have been measured, the final exponential term can be calculated for the contribution of each reflection hkl to each position xyz, but the phases of the reflections are unfortunately unknown, so the calculation cannot be carried out immediately.

The result of adding even just two waves varies from the sum to the difference of their amplitudes depending on their relative phases (Fig. 2.9), and to apply trial-and-error methods to thousands of waves is a task of impossible proportions.

We shall see in the next step (Section 2.8) that knowing part of the structure, i.e. the positions of some of the atoms, especially those with the most electrons, is often enough to help find the rest. The question is, where to begin?

Of the various methods used, two are by far the most common and important. One works best for structures containing one atom or a small number of atoms with significantly more electrons than the rest ('heavy atoms'), while the other is more appropriate for 'equal atom' structures, though in practice versions of it are used to solve most structures. In general, not surprisingly, the easiest atoms to find are those which contribute most to the total scattering.

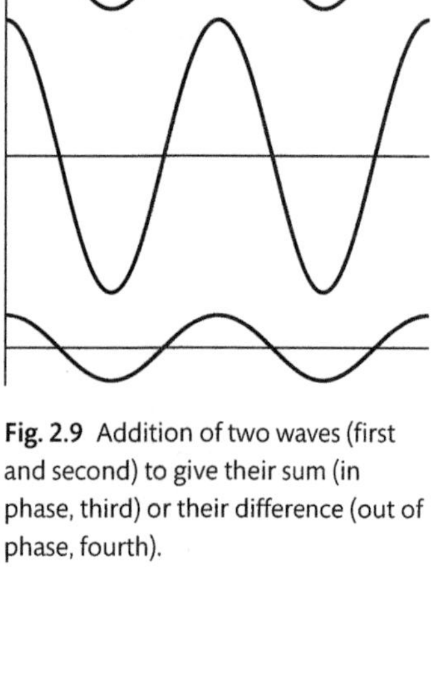

Fig. 2.9 Addition of two waves (first and second) to give their sum (in phase, third) or their difference (out of phase, fourth).

The Patterson synthesis

The Fourier transform of the observed diffracted beam amplitudes $|F_o|$ gives the correct electron density, but it requires knowledge of the phases of all the reflections (equation 1.20). The Fourier transform of the squared amplitudes F_o^2 with all phases set equal to zero (all waves taken in phase) produces what is called a **Patterson synthesis** (or Patterson function, or Patterson map). All the information needed for this transform is known; it can be calculated for any measured diffraction pattern. But is it of any use?

A. L. Patterson introduced and developed this method relatively early in the history of X-ray crystallography.

$$P(xyz) = \frac{1}{V} \sum_{h,k,l} |F_o(hkl)|^2 \cdot \exp[-2\pi i(hx + ky + lz)] \qquad (2.2)$$

The Patterson map looks rather like an electron density map (see Fig. 2.10), in that it has peaks of positive density in various positions. These are not, however, the positions of atoms in the structure. Instead, it turns out that the Patterson function is a map of vectors between pairs of atoms in the structure. For every pair of atoms at positions (x_1, y_1, z_1) and (x_2, y_2, z_2) there is a peak in the Patterson map at $(x_1 - x_2, y_1 - y_2, z_1 - z_2)$ and another of the same size at $(x_2 - x_1, y_2 - y_1, z_2 - z_1)$. In other words, for every peak seen in the Patterson map (at, say, u, v, w), there must be two atoms in the structure whose x coordinates differ by u, y coordinates differ by v, and z coordinates differ by w. The Patterson peaks show where atoms lie relative to each other, but not where they lie relative to the unit cell origin, which is what we really want to know.

Peaks in an electron density map are, ignoring vibration effects, proportional in size to the atomic numbers of the respective elements, since these are equal to the number of electrons. Patterson peaks are proportional in size to the product of the atomic numbers of the two atoms concerned.

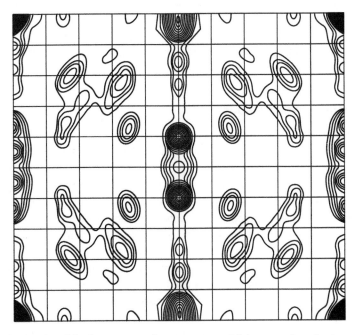

Fig. 2.10 A section of the Patterson map for a structure containing one unique As atom together with atoms of H, C, N, F, and S.

In order to see how the Patterson function can be used to locate some of the atoms, we note some of the properties of the Patterson function, which follow from its definition.

(a) Every atom forms a pair, and hence a vector, with every other atom, including with itself. So a unit cell containing n atoms gives n^2 vectors. Of these, the self-vectors (each atom to itself) have zero length and all coincide at the origin $(0, 0, 0)$. This is always the largest peak in any Patterson map. There are $n^2 - n$ other peaks.

(b) The vector between atom A and atom B is exactly equal and opposite to the vector between atom B and atom A. This means that a Patterson map always has an inversion centre, even if the crystal structure itself does not.

(c) Patterson peaks have a similar shape to electron density peaks, but are about twice as broad.

(d) As a consequence of points (a) and (c), there is usually considerable overlap of peaks, and not all will be resolved as separate identifiable maxima.

For these reasons, Patterson maps usually show large featureless regions of overlapped broad peaks, with significant peaks due to vectors involving 'heavy atoms'. If a structure contains only a few heavy atoms among a lot of lighter atoms, the Patterson map will show a small number of large peaks standing out clearly above the general background level.

In such cases it is usually possible to find a self-consistent set of atomic positions for the heavy atoms which explain the large Patterson peaks. Vectors between symmetry-related heavy atoms often lie in special positions, with some coordinates equal to 0 or ½, for example, and are easily recognized. Worked examples (and exercises for the reader) are given in Chapter 3. Solving a Patterson map is rather like a mathematical brain-teaser puzzle. Once the heaviest atoms have been found, the rest are located as shown in Section 2.8.

For example, for ferrocene $Fe(C_5H_5)_2$, Fe–Fe peaks, Fe–C peaks and C–C peaks have relative heights $26 \times 26 = 676$, $26 \times 6 = 156$, and $6 \times 6 = 36$; these are in the ratio approximately 19:4:1. Peaks involving H are even smaller. The few large Fe–Fe peaks (because there is more than one molecule per unit cell) will be clearly seen among the many smaller overlapped peaks. Figure 2.10 also shows a small number of large peaks and is otherwise relatively featureless.

Patterson search methods

Even for structures without particularly heavy atoms, the Patterson synthesis can provide a solution method in some cases. If a significant proportion of the molecule has a known shape, then a group of vectors generated internally by these atoms can be calculated. Such a pattern occurs in the Patterson map, but its orientation is unknown and it is mixed up with other vectors involving the rest of the molecule and vectors between atoms in different molecules. It may be possible to match the pattern by computer search and hence find the correct orientation of the known fragment. Further computer analysis of possible intermolecular vectors then gives proposed positions for the fragment. Steroids, with a relatively rigid and predictable tetracyclic nucleus, are good examples of suitable materials for this **Patterson search** approach (Fig. 2.11). **Direct methods** are, however, much more commonly used for 'equal atom' structures (and, together with **dual-space methods**, for most structures, as they are more automatic than Patterson-based methods).

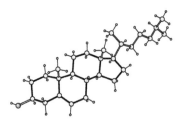

Fig. 2.11 The structure of cholesterol, with the characteristic rigid steroid tetracyclic framework highlighted.

Direct methods

This is a general name given to a wide variety of methods which seek to obtain approximate reflection phases from the measured intensities with no other information available. Such a description of the situation is, however, misleading. There is other information available to help us find the missing phases, in various aspects of the nature of the electron density we are trying to determine.

The electron density is the Fourier transform of the diffraction pattern. This means we add together a set of waves in order to produce the electron density distribution. Each wave has half its value positive and half negative (alternate 'crests' and 'troughs': see Fig. 1.21), except for $F(000)$, which is constant and positive. The electron density, however, is everywhere positive or zero; it can have no negative regions. Furthermore, it is concentrated into certain compact regions (atoms; see Fig. 1.5). So the waves must be added together in such a way as to build up and concentrate positive regions and cancel out negative regions. This puts considerable restrictions on the relationships among the phases of different reflections, especially the most intense ones, which contribute most to the sum.

Since large numbers of reflections are involved in the complete Fourier transform, individual phase relationships are not certainties, but have to be expressed in terms of probabilities, and the probabilities depend on the relative intensities.

Direct methods involve selecting the most important reflections (those which contribute most to the Fourier transform), working out the probable relationships among their phases, then trying different possible phases to see how well the probability relationships are satisfied. For the most promising combinations (assessed by various numerical measures), Fourier transforms are calculated from the observed amplitudes and trial phases, and are examined for recognizable molecular features.

Over the years, various methods of more or less sophistication have been developed for the steps involved. They can be regarded most simply as a sort of inspired trial-and-error method, in which it is usually necessary to try many different sets of phases and use the relationships themselves to 'refine' or improve them. Direct methods involve a considerable amount of computing, and are treated as a 'black box' even by many of their regular users. When they are successful, they usually locate most or all of the non-hydrogen atoms in a structure. Examples are given in the next chapter, and an illustration in one dimension is shown later in this section, after the description of other methods. Although starting sets of phases can be assigned by various theoretical approaches, it has been found that completely random values can lead to success in many cases if sufficient attempts are made (this is no problem with modern computers) and if effective procedures are used for refining phases on the basis of expected phase relationships.

Dual-space methods

Crystallographers often refer to a physical crystal structure as being in direct space (based on the direct lattice) and the diffraction pattern in reciprocal space

Direct space is measured in distance units, usually Å; the units of reciprocal space are Å^{-1}, as can be seen from the rearranged form of the Bragg equation (1.4). Fourier transforms relate functions with reciprocal or inverse units such as these. Other examples in chemistry include Fourier transform NMR spectroscopy (converting decay measurements in time to spectra in frequency, which has units of reciprocal time) and Fourier transform IR spectroscopy (converting interferometry measurements in cm to spectra in wavenumbers (cm^{-1})).

An online demonstration of the charge flipping procedure, as an example of dual-space methods, can be found at http://escher.epfl.ch/flip/

(based on the reciprocal lattice, which describes its geometry and is evident, for example, in Figs. 1.13, 2.5(b), and 2.7). The two spaces are related by forward and reverse Fourier transforms. Solving a structure by Patterson methods involves transforming the measured intensities into a Patterson function, which is in direct space, and then deducing atom positions from this function; this is thus a direct-space method. Conventional direct methods do most of their work in reciprocal space, manipulating reflection phases, with a Fourier transform into direct space to give a trial electron density map only after a promising set of phases has been found. Powerful methods of solving structures have been developed, combining alternating direct space and reciprocal space calculations, exploiting the available information and expected behaviour in each step. The various methods differ in the particular direct and reciprocal space techniques they use; all of them rely on fast repeated interconversion between the two spaces by Fourier transforms.

The starting point for dual-space methods may be in either direct or reciprocal space. A direct space starting model consists of some kind of electron density distribution or an initial set of atom positions. Examples include: a random allocation of expected atom types in the unit cell, possibly screened to avoid impossibly short contacts or other undesirable features; a molecular fragment of known geometry in random positions and orientations, or in positions suggested by an inconclusive Patterson search procedure; a structure of a related compound; or an electron density function calculated by special manipulations of the Patterson function. A reciprocal space starting model is a set of phases for some or all of the reflections: randomly generated, taken from another related structure, or calculated by other methods. Adjustments made to the reciprocal space information during each cycle of dual-space calculations include applying probable phase relationships to modify the phases, or simply taking the calculated phases from the forward Fourier transform and using them unaltered in combination with the observed amplitudes (ignoring the calculated amplitudes) in the subsequent reverse transformation. Modifications to the direct space electron density or atom-based model may include selecting atoms that give reasonable molecular geometry and ignoring others, moving or deleting atoms to improve the geometry, omitting a particular fraction of the atoms (these atoms being chosen at random), and changing the sign of all electron density below a particular threshold value (this method is known as **charge flipping** and has proved remarkably successful for such a simple idea; it has grown in popularity recently). Progress of the calculations is monitored by a range of 'figures of merit' that measure the agreement of either electron density or reflection phases (or both) with some kind of expected behaviour. In many cases, a recognizable molecular structure emerges, either gradually or suddenly, from the starting point. Some dual-space methods ignore the proposed space-group symmetry, carry out all calculations in space group $P1$ with only translation symmetry, and then locate the true symmetry elements of the structure in the electron density map (direct space) or by inspection of phase relationships for potentially symmetry-equivalent reflections (reciprocal space).

Other methods

Almost all 'small molecule' crystal structures these days are solved by Patterson, direct, or dual-space methods. Other methods used for macromolecular structures are described briefly in Chapter 4. A partial solution for some small molecule structures can be found from considerations of symmetry. For example, if a molecule which could reasonably have a centrosymmetric geometry crystallizes in the triclinic system with one molecule per unit cell, then the centre of the molecule probably coincides with a crystallographic inversion centre in space group P$\bar{1}$ (see Section 1.6). Such a situation is frequently found for metal complexes, and the metal atom is thereby located at a special position. If the metal has sufficient electrons to be classed as a heavy atom, it can be used, as shown in the next section, to find the rest of the atoms; no Patterson map or direct methods calculation is necessary. An extreme example, that of [Cr(NH$_3$)$_6$][HgCl$_5$] in a cubic space group with most of the atoms in special positions, was described in Section 1.6 with respect to symmetry arguments.

There is no 'correct' method for solving a particular structure. Once the right solution has been found, by whatever method, it can be further refined (see Section 2.9); the method of solution is no longer important. If one method does not work, it is perfectly valid to try others, with all the available variations, until one is successful. The objective is to beat the 'phase problem'; exactly how this is done does not really matter.

A one-dimensional illustration of direct methods and Patterson synthesis

Racemic 3-bromo-octadecanoic acid (Fig. 2.12) forms triclinic crystals with two molecules in the unit cell, related to each other by inversion symmetry. The unit cell has two short and one long axis, and the molecule is extended approximately along the longest axis (c) (Fig. 2.13). This structure provides an illustration of structure solution in one dimension; if we take only those reflections 00l which have h and k indices equal to zero, they contain no information about the x and y coordinates of atoms, but we can use them to find z coordinates. The reflections 001 and 002 were not measured, probably because they lie too close to the direct beam; intensities were obtained for reflections with l from 3 to 21, and their amplitudes, derived from the measured intensities, are in Table 2.1.

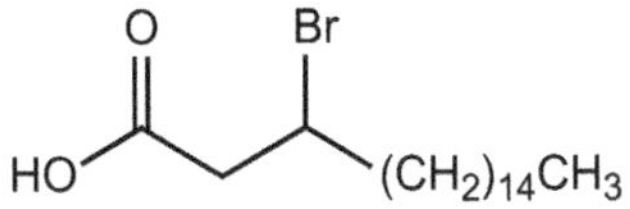

Fig. 2.12 The formula of 3-bromo-octadecanoic acid.

Fig. 2.13 Two molecules in the elongated unit cell of 3-bromo-octadecanoic acid.

To find the z coordinates of the atoms, it is necessary to add up the contributions of all these 19 waves, together with F(000), at each of a range of z values from 0 to 1 (because of the inversion symmetry in the structure, the range 0.5 to 1.0 is actually equivalent to the range 0.0 to 0.5 by inversion, but we shall carry out the calculations for the whole z range of one unit cell for completeness). For this purpose we need a one-dimensional version of equation 1.20.

$$\rho(z) = \frac{1}{c}\sum_l |F(l)| \cdot \exp[i\phi(l)] \cdot \exp[-2\pi i(lz)] \tag{2.3}$$

At each chosen value of z, there are 19 terms to add together. The task is further simplified by the fact that the structure is centrosymmetric (this simplification for centrosymmetric structures applies also in three dimensions). In such cases the phases of reflections can only take one of two special values, 0° and 180° (or 0 and π radians), and the term exp(iφ) becomes equal to cos(φ); this is simply +1 for φ = 0°, and −1 for φ = 180°, and the mystery of the unknown phase narrows

Table 2.1 Observed amplitudes and correct phases for 00*l* reflections of 3-bromo-octadecanoic acid

l index	Measured |F(00*l*)|	Correct sign
3	5.8	+
4	45.2	–
5	39.2	–
6	52.6	–
7	10.6	–
8	3.8	+
9	32.2	+
10	31.8	+
11	30.4	+
12	11.8	+
13	6.2	–
14	18.2	–
15	21.8	–
16	16.2	–
17	8.2	–
18	10.0	+
19	14.4	+
20	23.4	+
21	44.6	+

down to a straight choice between completely in phase and completely out of phase for each individual reflection. This is the same as having to choose whether each amplitude is added or subtracted to make up the total sum. This still leaves us with a two-way choice 19 times over, giving 524 288 possible combinations! This is clearly too much for a blind trial-and-error approach.

Note also that the final exponential term simplifies to a cosine in the same way, so we have the equation

$$\rho(z) = \frac{1}{c}\sum_{l}|F(l)| \cdot \text{sign}[F(l)] \cdot \cos[2\pi(lz)] \tag{2.4}$$

For the one-dimensional case, this summation process can be shown graphically. The contributions of each of the 19 reflection amplitudes to each point across the range of z from 0 to 1 are shown in Fig. 2.14. The contributions are all shown on the same scale, and with all the unknown signs (corresponding to the unknown phases) set arbitrarily as positive. The cosine term in equation 2.4 means that reflections with a low value of the index l make a contribution to the electron density which varies only slowly across the unit cell; reflections with a high value of l contribute much more finely, with more maxima and minima across the range. Figure 2.15 shows the result of adding up these 19 contributions with different combinations of signs (phases). The top result comes from arbitrarily chosen signs; it does not look like a promising solution for the electron

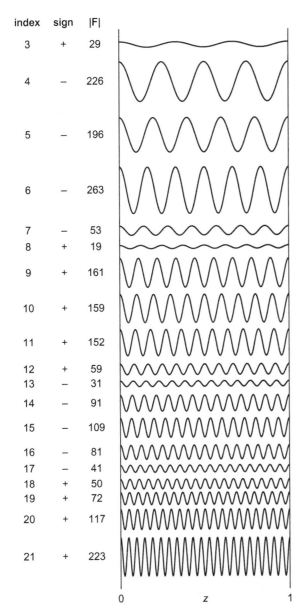

index	sign	|F|
3	+	29
4	−	226
5	−	196
6	−	263
7	−	53
8	+	19
9	+	161
10	+	159
11	+	152
12	+	59
13	−	31
14	−	91
15	−	109
16	−	81
17	−	41
18	+	50
19	+	72
20	+	117
21	+	223

0   z   1

Fig. 2.14 The contributions of the 19 00l reflections to the one-dimensional Fourier summation of equation 2.3, with all their phases set at zero. The correct phases, as signs, are shown with the indices and amplitudes in the left-hand columns. Reflections with larger indices are observed at higher Bragg angles and provide greater resolution of the electron density image, just as light scattered at greater angles by an object on an optical microscope provides better resolution than low-angle scattering. The curves shown here for the different reflections must not be confused with X-ray wavelengths and frequencies (the wavelength is constant for all reflections); these are not the waves themselves, but the contributions they make to the electron density calculation at various points in the unit cell via the one-dimensional Fourier transformation.

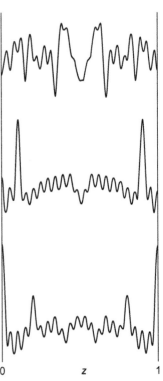

0   z   1

Fig. 2.15 Combinations of the 19 contributions of Fig. 2.14 with different sets of phases: top, randomly chosen phases giving an unrecognizable result; middle, correct phases clearly showing the bromine atoms; bottom, all phases positive, resembling a Patterson synthesis.

density, particularly with some deep minima. The second result comes from the correct signs (as provided by a forward Fourier transform calculation from the final known structure; these correct signs are given in Table 2.1); it shows large maxima for two symmetry-related bromine atoms (two molecules per unit cell) and smaller maxima, approximately regularly spaced, most of which correspond to pairs of carbon atoms, these overlapping in projection along the axis. The positions of the atoms, particularly the large bromine atoms, are correctly given: the final refined z coordinates for bromine are very close to 0.1 and 0.9.

To illustrate the principles of direct methods, concentrate on the largest amplitudes; clearly these contribute most to the summations, and incorrect signs for the smaller amplitudes will not greatly affect the result. The correct signs for reflections 4 and 5 (using the l indices to label them) are both negative. This means they should both be turned upside down before being added into the sum. Together they then contribute a considerable positive amount where their original first and last troughs, as seen in Fig. 2.14, almost coincide, a large negative amount at $z = 0$ and $z = 1$, and their contributions largely cancel out elsewhere. Reflection 9 is also quite strong, and the positions of its crests and troughs are related to those of reflections 4 and 5, simply because $4 + 5 = 9$. If reflection 9 is to reinforce the positive build-up of electron density provided by the inverted reflections 4 and 5, rather than partially cancelling it, it must have a crest roughly coinciding with the first troughs of those reflections as they are shown in the figure, so its sign must be positive and not negative. Since $(-1) \times (-1) = (+1)$ this relationship can be represented as

$$\text{sign}[F(9)] = \text{sign}[F(4)] \times \text{sign}[F(5)] \tag{2.5}$$

or, in terms of phase angles rather than signs,

$$\phi(9) = \phi(4) + \phi(5) \tag{2.6}$$

The probability that such a relationship among the phases of reflections with related indices is true increases with their amplitudes. In this particular example, if we take all the reflections with amplitudes greater than 80, there are 11 of these relatively strong reflections and 19 relationships of this kind which involve only these reflections. All 19 relationships are, in fact, obeyed (for example, reflection 6 is negative, reflection 9 is positive, and reflection 15 is negative: $(-1) \times (+1) = (-1)$); Table 2.2 gives them all. In general, for three-dimensional structures, because these relationships are probabilities rather than certainties, some of the indications will be wrong, and for some reflections participating in several relationships contrary indications may be found, so an overall balance of the various indications has to be taken.

In direct methods as commonly applied, the strongest reflections are chosen and all the phase relationships among them are generated. Then various possible combinations of phases are tried, either by assigning values to a few reflections and using the probability relationships to generate others, or by assigning random phases to all the reflections and using the relationships to improve them so that they fit the relationships better. The combination of phases giving the best agreement with the expected relationships is then used, together with the

It is not actually the strongest reflections absolutely which are most important, but those reflections which are relatively strong for their Bragg angle; because intensities always decrease at higher angle, a high-angle reflection which is weak compared with low-angle reflections but strong compared with other high-angle reflections is an important contributor to the electron density calculation. For this reason, direct methods use *normalized structure amplitudes*, more commonly known as **E values**, instead of the basic amplitudes (*F values*) themselves; normalization is a calculation which effectively compensates for the θ-dependent decrease in intensity within a data set.

Table 2.2 Phase relationships for the strongest 00*l* reflections

	4	5	6	9	10	11	14	15	16	20	21
<u>4</u>		9	10		<u>14</u>	<u>15</u>			20		
<u>5</u>	9		11	<u>14</u>	<u>15</u>	<u>16</u>		20	21		
<u>6</u>	10	11		<u>15</u>	<u>16</u>		20	21			
9		<u>14</u>	<u>15</u>			20					
10	<u>14</u>	<u>15</u>	<u>16</u>		20	21					
11	<u>15</u>	<u>16</u>		20	21						
<u>14</u>			20								
<u>15</u>		20	21								
<u>16</u>	20	21									
20											
21											

For each relationship the *l* indices of the three reflections are given by one entry in the table body together with the corresponding number at the head of the column and the number at the left-hand end of the row; an underlined index represents a negative reflection amplitude (for example, reflection 6 has a negative amplitude, but reflection 9 has a positive amplitude).

observed amplitudes, in a reverse Fourier transformation to calculate an electron density map, and this is examined for recognizable molecular features. In typical cases a few tens of initial phase sets are tried and several of these are likely to lead to a recognizable correct structure showing most or all of the non-hydrogen atoms. In more difficult cases, many hundreds or thousands of attempts may be necessary.

The third result in Fig. 2.15 comes from taking all the reflections as positive, i.e. all phases set at zero; the 19 curves of Fig. 2.14 are simply added together as they are. Since all the curves have a crest at $z = 0$ and at $z = 1$, very large maxima are generated at these points. This is just like a one-dimensional Patterson synthesis, except that $|F|$ values have been used instead of F^2 values. The corresponding result using F^2 and zero phases looks very much like the third curve except that the maxima and minima are more exaggerated. The large maximum at the origin is a feature of all Patterson syntheses, corresponding to the superposition of the self-vectors of all atoms in the structure. The next largest maximum in this curve (together with its symmetry equivalent) corresponds to a vector between the two bromine atoms in the unit cell; its z coordinate is twice the z coordinate for one bromine atom, because the two symmetry-equivalent bromine atoms are at $+z$ and $-z$, and the difference between these is $(+z) - (-z) = 2z$. Hence the bromine atom in the molecule can be located simply by inspection of this Patterson synthesis, without the knowledge or guessing of any phases at all. Location of the remaining atoms then follows in the next stage (Section 2.8). The solution of a Patterson synthesis in order to find one unique heavy atom in a structure (together with its symmetry equivalents) often involves no arithmetic more difficult than dividing by 2. Further worked examples are given in Chapter 3.

This structure has been published. See: The crystal and molecular structure of DL-3-bromo-octadecanoic acid. S. Abrahamsson and M. M. Harding, *Acta Crystallogr.* 1966, 20, 377–383; the CSD REFCODE is BRODAC.

This one-dimensional Fourier synthesis is available as a Microsoft Excel spreadsheet for use by the reader in exploring various features of Fourier calculations including also the structure completion steps described in the next section. For a description of the spreadsheet and its use, see: An Excel spreadsheet for a one-dimensional Fourier map in X-ray crystallography. W. Clegg, *J. Chem. Ed.* 2004, **81**, 908–911.

2.8 Completing the basic structure

If the initial structure solution has revealed positions for all the atoms (except for hydrogen atoms, which have very little electron density and are not usually found until later, if at all), this next step is unnecessary. Often, however, particularly from analysis of a Patterson map or for structures containing solvent molecules or other components with high atomic vibration or disorder, only a partial structure has been obtained: some atom positions are known, but not all. This partial structure serves as our initial model or **trial structure**.

Using the forward Fourier transform equation (the mathematical representation of the scattering process), we can calculate what the diffraction pattern would be if this model structure were, in fact, the correct complete structure:

$$\text{model structure} \xrightarrow{\text{FT}} \text{set of } F_c \qquad (2.7)$$

where F_c are calculated structure factors, one corresponding to each observed structure factor F_o. The calculation provides values for the amplitudes and phases of F_c ($|F_c|$ and ϕ), whereas we have only amplitudes for F_o ($|F_o|$, no ϕ).

If the atoms of the model structure are approximately in the right positions, there should be at least some degree of resemblance between the calculated diffraction pattern and the observed one, i.e. between the sets of $|F_c|$ and $|F_o|$ values. The two sets of values can be compared in various ways. The most widely used assessment is a so-called *residual factor* or **R factor**, defined as

$$R1 = \frac{\sum \left| \, |F_o| - |F_c| \, \right|}{\sum |F_o|} \qquad (2.8)$$

This involves adding together all the discrepancies between corresponding observed and calculated amplitudes, ignoring signs of the differences, and normalizing the sum by dividing by the sum of all the observed amplitudes to give a value which can be compared for different structures. Variations on this definition include using F^2 values instead of $|F|$ values, squaring the differences, and/or incorporating different weighting factors (**weights**) multiplying different reflections, based on their s.u.s, and hence incorporating information on the relative reliability of different measurements; for example, one residual factor in a very widely used computer program for crystal structure determination is

$$wR2 = \sqrt{\frac{\sum w(F_o^2 - F_c^2)^2}{\sum w(F_o^2)^2}} \qquad (2.9)$$

where each reflection has its own weight w. This is, in many ways, and certainly from a statistical viewpoint, more meaningful than the basic $R1$ factor. For a correct and complete crystal structure determined from well-measured data, $R1$ is typically around 0.02–0.07; for an initial model structure it will be much higher, possibly 0.4–0.5 depending on the fraction of electron density so far found, and its decrease during the next stages is a measure of progress. Values of $wR2$

and other residual factors based on F^2 are generally higher than those based on F values, by a factor of two or more.

Obviously, the reverse Fourier transform (the mathematical representation of the image construction) carried out with the calculated amplitudes $|F_c|$ and calculated phases ϕ_c would just regenerate the electron density of the model structure,

$$|F_c| \text{ with } \phi_c \xrightarrow{\text{FT}^{-1}} \rho \text{ for model structure} \qquad (2.10)$$

and this is not progress. However, combination of the experimentally observed amplitudes $|F_o|$ (which carry information about the true structure) with the calculated phases ϕ_c (which are not completely correct, but are the best approximation we currently have to the unavailable ϕ_o values) produces something new:

$$|F_o| \text{ with } \phi_c \xrightarrow{\text{FT}^{-1}} \rho \text{ for a new model structure} \qquad (2.11)$$

Usually, if the errors in the calculated phases are not too large, this electron density shows the atoms of the existing model structure, together with additional atoms not already known. This provides an improved model structure, with more atoms than before.

If there are still more atoms to be found, this process can be repeated. A forward Fourier transform of the new model structure gives a new set of $|F_c|$ and ϕ_c; the previous set is discarded. The new $|F_c|$ and the unchanged original $|F_o|$ values should now give lower R factors, and the improved ϕ_c together with $|F_o|$ generate, via another reverse Fourier transform, a further electron density map.

Eventually all the atoms are located and the Fourier transform calculations give no further improvement. This repeated process is an example of a *bootstrap* procedure. It is also another example of a dual-space method, the direct space manipulation being the selection of genuine atoms from the candidate electron density peaks, and the reciprocal space manipulation being the combination of observed amplitudes with phases calculated from the model structure.

There are some variations on the basic Fourier bootstrap procedure, which make it more effective. In particular, the reverse transform (Fourier map calculation) can be carried out using the differences $|F_o|-|F_c|$ instead of just $|F_o|$. In this case, a **difference electron density map** is produced, in which the existing atoms of the current model structure do not appear. This makes new atoms stand out more clearly from the background and from false maxima arising from errors in the ϕ_c values. Difference electron density peaks or holes (negative peaks) at model structure atom positions may indicate incorrect atom assignments with too little or too much assumed electron density, which should be corrected in the next model.

Full electron density maps, with values of electron density or difference electron density at each of many regularly spaced points in all or part of the unit cell, either numerically or as contours, are not often generated and output as such by computer programs used for chemical crystallography. In most cases an automatic search is carried out for the positions of maximum density (peaks, analogous to mountains on a geographical map), and the output is just a list of these in descending order of height, as potential positions of atoms.

$wR2$ is the conventional name for this residual factor. The w indicates that weights are included, and the 2 indicates that F^2 values are used rather than F values; compare this with $R1$ defined above (which does not include weights and is based on F values; historically this was usually referred to as just R).

'Bootstrap' is a term common in computing jargon, usually shortened to 'boot', and is derived from an old saying, 'You can't pull yourself up by your own bootstraps' i.e. shoe-laces; in computer operating systems, crystallography and many other sciences, you can!

There are many other variations on the straightforward Fourier map calculation, using combinations of F_o, F_c, and normalized E_o values, and applying weighting schemes designed to enhance the contributions of some reflections (perhaps those for which the calculated phases are likely to be more nearly correct) and reduce the contributions of others. They serve particular purposes at different stages of a difficult structure solution.

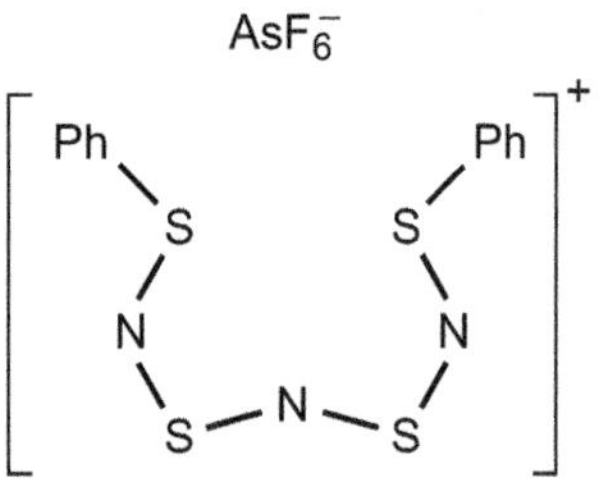

AsF_6^-

Fig. 2.16 The component ions of [PhSNSNSNSPh]$^+$ [AsF$_6$]$^-$.

An example of structure completion by Fourier syntheses

To illustrate the bootstrap procedure we take a structure which has most of the atoms lying in one plane, because the gradual development of the structure is clearly seen by successive calculations of the electron density in this plane. The compound is [PhSNSNSNSPh]$^+$[AsF$_6$]$^-$ (Fig. 2.16). The complete cation, together with arsenic and two fluorine atoms of the anion, lie on a mirror plane perpendicular to the b axis in an orthorhombic space group, so all their y coordinates are equal. The position of the arsenic atom within this plane can be found by inspection of the Patterson synthesis, part of which was shown in Fig. 2.10 (As has 34 electrons, S 16, F 9, N 7, and C 6, so vectors between symmetry-equivalent arsenic atoms stand out clearly).

Calculation of the diffraction pattern corresponding to the arsenic atom alone (and its equivalent atoms according to the space-group symmetry), by the forward Fourier transform equation (2.7), gives a set of calculated amplitudes and phases for this very crude initial model structure; at this stage the value of $R1$ is 0.604, so the agreement between the observed and calculated amplitudes is not very good. The calculated phases are also far from perfect (the structure is centrosymmetric, so each phase must be either 0° or 180°, and any particular calculated phase is either completely right or completely wrong), but there are enough correct phases for the reverse Fourier transform, calculated from these phases and the observed amplitudes (equation 2.11), to show not only the known arsenic atom, but also four clear peaks for the four sulfur atoms (Fig. 2.17(a)). Inclusion of these in the model structure leads to a calculated diffraction pattern in better agreement with the observed amplitudes: $R1$ is reduced to 0.364, the calculated phases are more nearly correct, and the next electron density map (reverse Fourier transform) shows all the N, C, and F atoms (Fig. 2.17(b)). With all the non-hydrogen atoms included, $R1$ drops to 0.036 after refinement (see Section 2.9), and an electron density map shows well resolved and clear peaks for all the atoms (Fig. 2.17(c)). The largest peaks in a difference electron density map (calculated from $|F_o|-|F_c|$ instead of from $|F_o|$) are in the positions expected for the hydrogen atoms (Fig. 2.17(d)), and incorporation of these into the model structure leads to a final $R1$ of 0.027, and a very precise structure.

2.9 Refining the structure

Once all the non-hydrogen atoms have been found, the model structure needs to be subject to **refinement**. This means varying the numerical parameters describing the structure to produce the 'best' agreement between the diffraction pattern calculated from it by a Fourier transform and the observed diffraction pattern. Since there are no observed phases, the comparison of observed and calculated diffraction patterns is made entirely on their amplitudes $|F_o|$ and $|F_c|$. Changing any of the structural parameters (modifying the model structure in any way) affects the $|F_c|$ values, while the $|F_o|$ values remain fixed during the process.

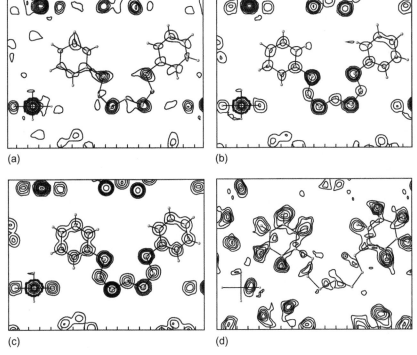

(a)   (b)

(c)   (d)

This structure has been published. See: Reaction of [SNS][AsF$_6$] with Hg(CN)$_2$ and PhHgCN: preparation and crystal structures of [Hg(CNSNS)$_2$][AsF$_6$]$_2$ and [PhS$_4$N$_3$Ph][AsF$_6$]. C. M. Aherne, A. J. Banister, I. Lavender, S. E. Lawrence, J. M. Rawson and W. Clegg, *Polyhedron* 1996, **15**, 1877 -1866; the CSD REFCODE is TEQSOX.

Fig. 2.17 Successive electron density and difference electron density syntheses in the development of the structure of [PhSNSNSNSPh]$^+$[AsF$_6$]$^-$ starting from the position of the arsenic atom derived from a Patterson map. The contour interval is much smaller for the difference electron density map (d).

The refinement process uses a well-established mathematical procedure called *least-squares* analysis, which defines the 'best fit' of two sets of data (here $|F_o|$ and $|F_c|$) to be that which minimizes one of the least-squares sums:

$$\sum w(|F_o|-|F_c|)^2$$
$$\text{or } \sum w(F_o^2 - F_c^2)^2$$

(2.12)

The first of these (refinement on F) has historically been most commonly used, but the second (refinement on F^2) is now regarded as standard and is, in many ways, superior. The contribution of each reflection to the sum is weighted according to its perceived reliability, usually with weights based on the experimental s.u.s, such as $w=1/\sigma^2(F_o^2)$ for refinement on F^2.

The least-squares refinement of crystal structures is similar, in principle, to finding a 'best-fit' straight line through a set of points on a graph, but is more complicated because (i) there are many variable parameters instead of just two (the gradient and intercept) for a straight line graph and (ii) the equation relating data to parameters (the Fourier transform) is far from linear. Because of the non-linearity, an approximate solution (the model structure) must be known before refinement can begin, and each least-squares calculation is approximate, not

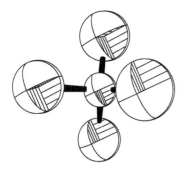

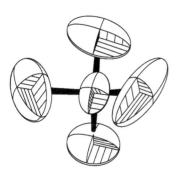

Fig. 2.18 Isotropic (top, represented as spheres) and **anisotropic** (bottom, represented as ellipsoids) atomic displacements for a perchlorate anion.

As a rough rule of thumb, around 100 data per non-hydrogen atom in the asymmetric unit should be ample.

The **goodness of fit** is another standard statistical parameter, intended to show how well the calculated diffraction pattern corresponding to the model structure agrees with the observed diffraction pattern. For an ideal agreement and a correct weighting scheme, the goodness of fit should have a value of unity; considerable variation is observed in practice.

exact, giving an improvement to the model, but not the best possible fit; the calculation must be repeated several times until eventually the changes in the parameters are insignificant.

What are the numerical parameters to be refined? They are, for the most part, the terms describing the positions and vibrations of the atoms in the Fourier transform equation (1.19). For each atom there are three positional coordinates x, y, z and a displacement parameter U, which can be interpreted as an *isotropic mean-square amplitude of vibration* (in Å^2) of the atom. In most experiments a significantly better fit to the data can be achieved by using more than one displacement parameter per atom in the model structure, allowing each atom to vibrate by different amounts in different directions (*anisotropic vibration*). The usual mathematical treatment has six U values (one for each axis and three cross-terms) for each atom in order to give different vibration amplitudes in three orthogonal directions which are, in general, not along the unit cell axes (Fig. 2.18). Thus, there are commonly nine refined parameters for each independent atom (atoms which are not related to each other by symmetry) in the structure. In addition, a scale factor has to be refined, which puts the $|F_o|$ and $|F_c|$ values on the same scale (the $|F_o|$ scale is arbitrary at the time of measurement, but $|F_c|$ values are calculated relative to the scattering power of one electron). There may be a small number of other refined correction factors, but most of these are not important for a basic understanding of the procedure.

Although there are many parameters to be refined for all but the smallest structures, the diffraction experiment usually provides an even greater number of observed data, unless X-ray scattering is unusually weak. Typically, the data/parameter ratio exceeds 6, and it may be as high as 20 or more. The structure refinement problem is said to be 'over-determined', and this is essential in order to produce precise (reliable) parameters. As well as providing a value for each refined parameter, the least-squares process also gives a standard uncertainty. These parameter s.u.s depend on the s.u.s of the data (a good structure requires good data!), on the extent of agreement of the observed and calculated data (a lower least-squares sum gives lower parameter s.u.s; another function closely related to this sum is called the **goodness of fit**), and on the excess of data over parameters (a greater excess gives lower parameter s.u.s). Both the quality and the quantity of measured data matter for the quality (reliability) of the structure derived from them.

Once the model structure has been refined with **anisotropic displacement parameters** for the atoms, it is often possible to see small but significant difference electron density peaks in positions close to those expected for hydrogen atoms, particularly if there are few or no heavy atoms in the structure. Hydrogen atoms are more likely to be located from measurements taken at low temperature, because this reduces the vibration of the atoms and sharpens the electron density peaks. It is possible to include the hydrogen atoms in the refinement, and this may improve the fit slightly, but their parameters usually have large s.u.s because their low electron density means they contribute only weakly to the diffraction of X-rays, so the measured intensities are relatively insensitive to the hydrogen atom parameters. In most cases, refinement is more successful if **constraints** are applied to hydrogen atom parameters, e.g. by keeping their bond lengths fixed and by tying their U values to those of the atoms to which they are attached. Details of how this is done vary enormously with different computer programs and with the habits and preferences of different crystallographers; some examples can be seen in Chapter 3.

In some cases, unconstrained parameters may refine to unreasonable values, or have unacceptably high uncertainties because they are not well defined by the diffraction data, but a rigid constraint would be inappropriate. A more flexible approach is the use of **restraints** (also known as soft constraints). Here a desirable target value is chosen for a particular parameter, or for a feature that can be calculated from a combination of parameters (such as a bond length or angle, planarity of a group of atoms, or relationships between displacement parameters of connected atoms), and the difference between this target and the value calculated from the current model is squared, appropriately weighted relative to other restraints and to the observed diffraction data, and added to the least-squares minimization function. The full set of restraints (of which there may be a few or many) thus contributes as effective 'observations' along with the diffracted X-ray amplitudes to control the refinement of the parameters of the model. This approach is particularly useful when diffraction is weak, leading to a shortage of significantly observed data.

The refinement stage usually involves the vast majority of the computing resources used in a crystal structure determination, simply because the calculations are many and very repetitive. Compared with finding the atoms initially, it is often a much less interesting process, but its correct execution is very important, since it delivers the final parameters describing the structure; exactly how a structure is solved does not really matter, but how it is refined does. At the end of refinement, a difference electron density map should contain no significant features (peaks or holes). This calculation is usually performed as an extra check on the validity of the refined model structure. Typically, a final map with no features outside the range ± 1 e Å^{-3} is accepted, along with other indicators, as evidence of a satisfactory structure determination.

Constraints are conditions which are imposed on the refinement, for example by requiring certain parameters to have particular values rather than being free to take values which give the best agreement between observed and calculated diffraction patterns. Constraints may be imposed for various reasons, including the requirements of symmetry or the need to control parameters which are poorly defined by the diffraction data.

Constraints are rigid mathematical relationships that must be obeyed in the refinement; they reduce the number of refined parameters while leaving the data untouched. **Restraints**, by contrast, are treated as non-diffraction 'experimental observations' and are combined with the diffraction data; they do not change the number or nature of the refined parameters. It is important to use an appropriate balance of weights for the restraints and the data so that both make a sensible contribution to the refinement; heavily weighted restraints behave almost like constraints.

2.10 Disorder, twinning, and the determination of 'absolute structure'

Disorder

An ideal crystal structure is completely ordered: each atom occupies a single well-defined site, all asymmetric units are exactly equivalent under the space-group symmetry, and all unit cells are identical. Instantaneously, of course, this is never true, as each atom is undergoing vibration and these movements are not usually correlated throughout the structure, but this effect is dealt with by the atomic displacement parameters, giving equivalence on a time-averaged basis. Large amplitudes of vibration are sometimes referred to as **dynamic disorder**, and they are reduced by cooling the sample in low-temperature data collection.

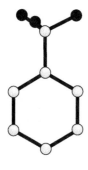

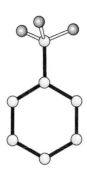

Fig. 2.19 A disordered trifluoromethyl group attached to a benzene ring: top and centre, the two observed disorder components; bottom, the combined disorder model seen in projection along the bond between the CF$_3$ group (in front) and the ring (behind).

Examples of constraints and restraints are given in Chapter 3, including application to disordered structures.

Some structures display **static disorder**, a random (not systematic) variation in the detailed contents of the asymmetric unit; here there are alternative positions for atoms or groups of atoms. If the disorder is truly random, then what X-ray diffraction sees is the average asymmetric unit. This appears in the model structure as partially occupied atom sites. It is best explained by giving some examples.

A commonly observed case is a methyl (CH$_3$) or trifluoromethyl (CF$_3$) group attached to an aromatic ring, as in a toluene solvent molecule or a tolyl substituent. There is no single preferred torsional orientation of the methyl group, and the energy barrier to rotation about the C–C bond is relatively low. In some structures, a difference electron density map will show three clear positions for the hydrogen or fluorine atoms, because the CH$_3$ or CF$_3$ group is held in one preferred orientation by neighbouring atoms in the same or another molecule. In others, six positions are found (with lower peak heights), corresponding to two alternative orientations with comparable energies, because the intermolecular interactions are weaker, and each molecule in the structure adopts one of the two possibilities at random, with no regard for the orientation adopted in neighbouring asymmetric units (Fig. 2.19).

Other commonly observed disorder patterns are for conformationally flexible groups of atoms, such as long alkyl chains, non-planar five-membered rings (tetrahydrofuran, THF, is notorious in this respect), counter-ions of high symmetry which are only loosely held in place with no strong intermolecular interactions (particularly BF$_4^-$, ClO$_4^-$ and PF$_6^-$ anions), and small solvent molecules not anchored by hydrogen bonding or other significant interactions (such as toluene, acetone, and dichloromethane). Figure 2.20 shows some typical cases. Further examples more relevant to inorganic compounds include substitutional disorder of two or more types of atoms or ions (e.g. a random distribution of Na$^+$ and K$^+$ cations over common sites in an alkali metal salt, or a random mixture of two different halide anions as ligands in a metal complex), and an end-to-end disorder of orientation of bridging cyano ligands in polymeric complexes containing M–C–N–M linkages.

Where the disordered atom sites are well resolved, so that individual electron density peaks can be seen, refinement is usually straightforward. When disordered atom sites are closer together than normal bonding distances, constraints and/or restraints may be needed, so that the expected molecular geometry is used as data in the refinement alongside the diffraction pattern intensities; this is particularly useful, and often essential, in cases of high disorder.

Very often, the disordered part of a crystal structure is not of particular interest, and the fact that this portion is less well determined is not important. Unfortunately, however, disorder in any part of the structure affects the reliability with which the whole structure can be determined. This is because the whole structure generally contributes to the whole diffraction pattern in the experiment, and the whole diffraction pattern contributes to the whole structure in the subsequent calculations; this is the nature of Fourier transforms. The effect can be seen in two particular ways.

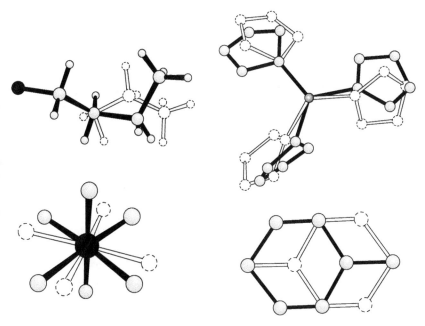

Fig. 2.20 Some examples of disorder: top left, an *n*-butyl chain in which the last two carbon atoms (with their hydrogen atoms) adopt two alternative positions; top right, three disordered THF ligands coordinated to a metal atom; bottom left, an AsF_6^- anion with two orientations related by rotation about one of the linear FAsF units; bottom right, a toluene solvent molecule disordered over an inversion centre (hydrogen atoms not shown).

First, except in the simplest cases, disorder can be difficult to incorporate in the model structure which is refined, especially when some alternative atom sites lie close together or when there are multiple disorder sites. A less than ideal model structure makes it more difficult to match the calculated and observed diffraction patterns and so leads to higher uncertainties in all the refined parameters than there would be for a fully ordered structure.

Second, static disorder represents an effective spreading out of electron density from ideal ordered positions, and this, like atomic vibrations, increases interference effects and hence reduces diffracted intensities, particularly at higher Bragg angles (see Section 1.7). Badly disordered structures give diffraction patterns in which the intensities fall off rapidly at higher angles; a lower proportion of reflections will be of significant intensity than for an ordered structure of similar scattering power. A shortage of high-angle data with significant intensity leads inevitably to a structure with lower precision, not only because there are fewer data. Inspection of the Bragg equation (1.3 and 1.4) shows that high scattering angle corresponds to small *d*-spacings, i.e. the **resolution** of closely spaced features in the structure. The maximum Bragg angle for which data are measured dictates the effective minimum resolution to which the structure can be determined. The effect of omitting higher-angle data is illustrated in Fig. 2.21, where electron density maps have been calculated from all data with $\theta < 25°$ ($d > 0.84$ Å) and from the data with $\theta < 15°$ ($d > 1.37$ Å).

$$d = \frac{\lambda}{2}\left(\frac{1}{\sin\theta}\right)$$

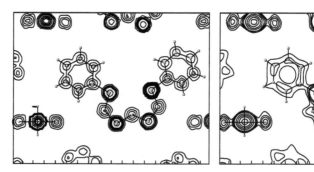

Fig. 2.21 Electron density maps calculated from data with θ < 25° (left) and from data with θ < 15° (right).

With the lower resolution data only, it is much more difficult to distinguish the individual lighter atoms.

High disorder particularly affects crystal structures of biological macromolecules, such as proteins, which incorporate large amounts of disordered solvent water in the substantial spaces between molecules. The diffraction intensities are also weak because of the large size of the molecules. As a consequence, usually only relatively low-angle intensities are observed, and atomic resolution is rarely achieved; in many cases, resolution is limited to 2 Å, 3 Å, or even worse. This is one of the challenges of **macromolecular crystallography** discussed further in Chapter 4. Similar, though less serious, problems can affect some large non-biological structures in popular modern research areas such as supramolecular chemistry. Low-temperature data collection and the use of synchrotron radiation are both important as means of maximizing intensities and crystal stability.

In some extreme cases of disorder, individual atom sites cannot be allocated from difference electron density maps, so there is no sensible contribution to the structural model for refinement; diffuse electron density is seen rather than discrete peaks. There are methods available that calculate a Fourier transform of this diffuse density in the structure and then treat this as an extra contribution (with amplitudes and phases) to be added to the diffraction pattern calculated from the rest of the structure, the sum of these being matched to the observed intensities in the refinement. The volume of the diffuse region and the number of electrons in it can be estimated as part of this calculation, and may be used, together with other information such as chemical analysis or spectroscopy, to identify the disordered component, which is usually a solvent, but this is not always successful. The main part of the structure is usually determined satisfactorily despite this problem. An example occurs in one of the case studies in Chapter 3.

Twinning

Another departure from ideal structures which can seriously hinder a crystal structure determination is the phenomenon of **twinning**. A twinned crystal is one in which two (or more) orientations or mirror images of the same structure occur together in a well-defined relationship to each other. It tends to occur when there are fortuitous rational relationships among the unit cell parameters, such as for a monoclinic structure with the angle β close to 90°, or with similar values for the a and c axis lengths. Twinning then results from 'mistakes' in putting the unit cells together to form the complete crystal during its growth, because they can fit almost equally well two different ways round.

A twinned crystal gives a diffraction pattern which is the superposition of the diffraction patterns of the two (or more) individual component parts of the crystal. In some cases, the two diffraction patterns have reflections which coincide exactly, each measured intensity then being the sum of two different (and non-equivalent by symmetry) but twin-related reflections. In other cases, the presence of two diffraction patterns can be recognized from the outset, because they are not exactly superimposed and the observed pattern clearly cannot belong to a single untwinned crystal. Whatever the precise nature of the twin relationship, if it can be worked out from the observed diffraction pattern (or from subsequent recognition of problems in the structure determination), then there are methods for solving and refining the structure, though it is more complicated than for a normal untwinned structure. A two-component twin is characterized by two parameters, which need to be determined: a **twin law**, which is expressed as a 3×3 matrix relating the orientations of the two components, and a **twin fraction** giving the relative amounts of the two components present in the crystal. An example of a twinned structure is included in the case studies of Chapter 3.

Absolute structure

In Chapter 1 it was shown with optical analogues how a diffraction pattern has inversion symmetry, even if the structure responsible for it does not. This is known as **Friedel's law**. In fact, this is only approximately true, because of an effect known as **resonant scattering**. As a first approximation, every time an atom scatters X-rays, a phase shift of 180° occurs; because this phase shift is constant, it can be ignored, and we regard all atoms as scattering in phase at $\theta = 0°$. In reality, the phase shift is not exactly 180°, and it is different for different atoms, generally increasing with atomic number, although there are irregularities in the pattern; atoms which contribute strongly to X-ray absorption also give significant resonant scattering. For centrosymmetric structures, the effects of resonant scattering on the pair of opposite reflections h, k, l and $-h, -k, -l$ are equal, and so they still do have the same intensity: Friedel's law is obeyed. For non-centrosymmetric structures, however, the effects do not cancel, and these reflections, known as *Friedel pairs* or *Friedel opposites*, have different intensities. The differences are not very great in most cases, since resonant scattering is only a small fraction of the total atomic scattering of X-rays, but careful measurement and comparison of Friedel pairs of reflections, or inclusion of them as separate non-equivalent data in refinement, allows us to distinguish a crystal structure from its inverse or opposite hand. For **chiral** molecules, this represents a direct experimental method of determining **absolute configurations**, which is not possible otherwise.

Resonant scattering is incorporated into the equations for X-ray diffraction by allowing atomic scattering factors to be complex numbers rather than purely

The conditions for possible twinning are actually more subtle than this and are often not obvious from simple inspection of the unit cell parameters. In geometrical terms, what is required is the possibility of taking either one unit cell or a block of two or more unit cells together and inserting this in a different orientation, and/or after reflection/inversion, into the same structure without significant distortion of the overall lattice of the rest of the structure. Apart from the geometrical aspect, it is also necessary for the differently oriented section of the structure to make acceptable energetic interactions with the 'host' structure. A full treatment of twinning requires a deeper knowledge of space-group and diffraction symmetry than is provided in this short text.

Twinning has been detected in many crystal structures since the widespread introduction of area-detector diffractometers, which record the whole diffraction pattern and not just the Bragg reflections found at positions expected from the initially found unit cell and crystal orientation. It is very likely that many earlier structures giving poor refinement results actually suffered from unrecognized twinning.

Friedel's law states that the intensities of reflections h, k, l and $-h, -k, -l$ have equal intensities; in fact they have the same amplitude, and phases with equal values but opposite signs:

$|F(h,k,l)| = |F(-h,-k,-l)|$, $\phi(h,k,l)$
$= -\phi(-h,-k,-l)$, and $I(h,k,l) = (I-h,-k,-l)$

Resonant scattering by an atom occurs when the X-ray photon energy is close to a value appropriate for promotion of an electron from one orbital of the atom to another, or for complete removal (ionization) of an electron. It is also often called **anomalous scattering,** but there is actually nothing anomalous about it, and the term resonant scattering relates the effect to the energy match that is responsible for it. The amount of resonant scattering depends on the element and on the X-ray wavelength.

A **chiral** molecule is one which is not identical to its mirror image; the two non-identical mirror images, known as **enantiomers**, are related like left and right hands. Determining the **absolute configuration** means finding out which one of the two we actually have.

real numbers; there are two extra contributions (measured, like the atomic scattering factor f itself, in units of electrons), one real and one imaginary (multiplied by i), so that f for each atom type is replaced by $f + f' + if''$. The values of f' and f'' are strongly wavelength-dependent, but do not depend on the Bragg angle θ. Some representative values are given in Table 2.3 for X-rays from the widely used copper and molybdenum targets.

Table 2.3 Normal and resonant scattering factors

Element	Normal $f(\theta = 0)$	f' and f'' (Cu)	f' and f'' (Mo)
Carbon	6	0.018, 0.009	0.003, 0.002
Oxygen	8	0.049, 0.032	0.011, 0.006
Phosphorus	15	0.296, 0.434	0.102, 0.094
Iron	26	−1.134, 3.197	0.346, 0.844
Iodine	53	−0.326, 6.836	−0.474, 1.812
Mercury	80	−4.292, 7.685	−2.389, 9.227

Although resonant scattering makes only a small contribution to the overall diffraction effects, in many cases this is sufficient to distinguish between a chiral structure and its opposite hand. One approach is to refine the two possible enantiomeric structures separately with the same set of experimental data, and decide which gives the better fit in terms of R factors and other indicators. A more common approach is to incorporate into the refinement a so-called *absolute structure parameter*, such that the structure is treated as a twin composed of the two enantiomers (the twin law is inversion) and the absolute structure parameter is the refined twin fraction, defined here as the fraction of the component of opposite hand to the model structure. An absolute structure parameter close to zero with a small s.u. therefore indicates that the model is correct, while a value close to 1 with a small s.u. shows that the model should be inverted. An intermediate value with a small s.u. suggests that the crystal really is an inversion twin, while a large s.u. demonstrates that the absolute structure cannot be reliably determined, usually because the resonant scattering contributions are too small. This approach has the advantage of providing, through the s.u. of the refined parameter, an estimate of the reliability of the result.

Absolute structure is a general term encompassing a number of physical properties that are different on inversion of a structure. Opposite chirality of two enantiomers is the property most familiar in molecular chemistry, but less familiar aspects such as crystal polarity are beyond the scope of this text.

2.11 Presenting and interpreting the results

What in fact are the results of a crystal structure determination? Returning to the microscope analogy, the application of the reverse Fourier transform equation to the observed diffraction pattern (but using calculated rather than genuinely observed phases!) gives an electron density map, an image of the X-ray scattering power of the crystal sample. It is, however, rare for the results to be presented in this way. Instead, the structure is represented as atoms (positioned at the centres of peaks of electron density) joined together by chemical bonds, and these atoms are described numerically by the refined parameters of the model structure.

The primary results from the refinement are the unit cell geometry and symmetry (space group), and the positions of all the atoms in the asymmetric unit (three coordinates each), together with their isotropic (one) or anisotropic (six) displacement parameters (each with an associated s.u.). The displacement parameters are usually interpreted as representing thermal vibration of the atoms and, in most cases, are regarded as less important and less interesting than the positional parameters; they are also more affected than the positional parameters by many experimental errors.

From the atomic coordinates, unit cell geometry and symmetry, many geometrical results can be derived. These include:

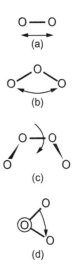

(a) **bond lengths** (distance between two atoms considered to be bonded together; see Fig. 2.22(a); a normal X-ray diffraction experiment does not directly show bonds, which are an interpretation based on distances and chemical experience);

(b) **bond angles** (angle between two bonds at one atom; see Fig. 2.22(b));

(c) **torsion angles** (the apparent angle between two bonds A–B and C–D when viewed along the B–C bond for a connected sequence of atoms A–B–C–D; see Fig. 2.22(c));

(d) the shapes and **conformations** of rings (e.g. chair and boat conformations of cyclohexane rings);

(e) the *planarity* or otherwise of groups of atoms (with possible consequences for the interpretation of their bonding);

(f) *degree of association* (monomers, formation of small **oligomers**, polymers);

(g) intermolecular geometry such as hydrogen bonding, van der Waals contacts, π-interaction stacking of planar aromatic groups.

Fig. 2.22 Geometrical parameters: (a) bond length; (b) bond angle; (c) torsion angle (two views, the second down the central bond).

As well as numerically, the results may be displayed graphically, as pictures of individual molecules and of the packing arrangement of molecules in the crystal structure (Fig. 2.23). Since these are all interpretative models, not direct observations (unlike what is seen through a standard optical microscope), a wide variety of styles of representation is possible, the traditional **ball-and-stick** model being the most commonly used. It is also possible, of course, to construct accurately scaled three-dimensional models of structures from the atomic coordinates, though this is likely to be very time-consuming or involve new 3D printing technology. In terms of the microscope analogy, the effective magnification for a typical molecule is $> 10^8$, which is a very impressive result!

Further interpretation and explanation of the structure and its relation to physical and chemical properties then follows as appropriate. For a large and complex structure, this can be quite a task. The amount and detail of structural information produced is greater than for any spectroscopic method of investigating chemical structure. It is salutary to recall that it all comes from a sample a fraction of a millimetre in size. Such is the power of the technique of X-ray crystallography.

This raises other questions, for example whether the particular crystal selected is actually representative of the bulk sample, which may not be a pure homogeneous compound. One way of checking this is to use **powder diffraction**, described in Chapter 4.

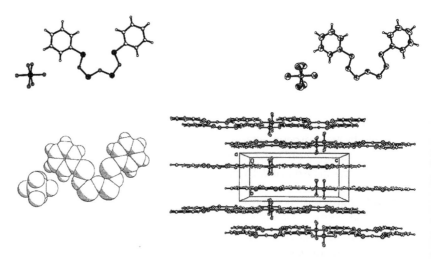

Fig. 2.23 Various styles of pictorial representation of the structure of $[PhSNSNSNSPh]^+[AsF_6]^-$: top left, conventional ball-and-stick model; top right, atomic displacement ellipsoids; bottom left, **space-filling** model; bottom right, packing of cations and anions in parallel layers in the crystal structure.

2.12 Archiving and reporting crystal structures

A crystal structure determination yields as its results the unit cell geometry and symmetry, and the positions and displacement (vibration) parameters of all the independent atoms it contains. From these the intramolecular and intermolecular geometry can be calculated, and graphical representations can be produced, as shown in the previous sections. All the results, together with the diffraction data from which they are derived, are held electronically in computer files.

The use of computers does not end with the successful completion of a structure refinement. One important further step is the safe storage of the results on computer-readable backup media locally and/or somewhere on the internet, for archiving, future access, and any further analysis.

Computers also play an important role in the publication of structural results in the research literature. Not only are manuscripts prepared with computer word-processors, as in research generally, but it is now normal practice to transmit results from researchers and authors to journals in purely electronic form, usually with procedures set up by the publishers. Such developments are very much assisted by the well-defined nature of the X-ray crystallographic results and widespread acceptance of particular standards and formats for them. A major development was the introduction of the so-called **Crystallographic Information File** (**CIF**), which was devised as a convenient and flexible form of information for archive, exchange, and publication; modern structure refinement programs generally produce a CIF as well as other forms of output, and further items of information can easily be added, since each piece of information in the file is uniquely identified by a name defined in an internationally agreed (and frequently updated) dictionary. An example of part of a CIF is shown in Fig. 2.24;

_cell_length_a	17.3546(11)
_cell_length_b	6.7359(7)
_cell_length_c	15.7608(9)
_cell_angle_alpha	90
_cell_angle_beta	90
_cell_angle_gamma	90
_cell_volume	1842.4(2)
_cell_formula_units_Z	4
_cell_measurement_temperature	240(2)
_cell_measurement_reflns_used	32
_cell_measurement_theta_min	10.59
_cell_measurement_theta_max	12.28
_exptl_crystal_description	needle
_exptl_crystal_colour	'dark blue'
_exptl_crystal_density_diffrn	1.851
_exptl_crystal_F_000	1016
_exptl_crystal_size_max	0.56
_exptl_crystal_size_mid	0.24
_exptl_crystal_size_min	0.16
_exptl_absorpt_coefficient_mu	2.358
_exptl_absorpt_correction_type	multi-scan
_exptl_absorpt_correction_T_min	0.342
_exptl_absorpt_correction_T_max	0.386
_exptl_absorpt_process_details	'based on equivalents and psi-scans'
_diffrn_ambient_temperature	240(2)
_diffrn_radiation_wavelength	0.71073
_diffrn_radiation_type	MoK\a
_diffrn_source	'sealed tube'
_diffrn_measurement_device_type	'Stoe-Siemens four-circle diffractometer'
_diffrn_measurement_method	'\w/\q scans with on-line profile fitting'
_diffrn_reflns_number	7487
_diffrn_reflns_av_unetl/netl	0.0318
_diffrn_reflns_av_R_equivalents	0.0506
_diffrn_reflns_limit_h_min	-20
_diffrn_reflns_limit_h_max	20
_diffrn_reflns_limit_k_min	-8
_diffrn_reflns_limit_k_max	8
_diffrn_reflns_limit_l_min	-18
_diffrn_reflns_limit_l_max	18
_diffrn_reflns_theta_min	1.745
_diffrn_reflns_theta_max	25.034
_diffrn_measured_fraction_theta_max	0.996
_reflns_number_total	1768
_reflns_number_gt	1444
_reflns_threshold_expression	'I > 2\s(I)'
_refine_ls_structure_factor_coef	Fsqd
_refine_ls_matrix_type	full
_refine_ls_weighting_scheme	calc
_refine_ls_weighting_details	
'w=1/[\s^2^(Fo^2^)+(0.0360P)^2^+0.8887P] where P=(Fo^2^+2Fc^2^)/3'	
_atom_sites_solution_primary	heavy
_atom_sites_solution_secondary	difmap
_atom_sites_solution_hydrogens	geom
_refine_ls_hydrogen_treatment	constr
_refine_ls_extinction_method	SHELXL
_refine_ls_extinction_coef	0.0017(3)
_refine_ls_extinction_expression	
'Fc^*^=kFc[1+0.001xFc^2^\l^3^/sin(2\q)]^-1/4^'	
_refine_ls_number_reflns	1768
_refine_ls_number_parameters	152
_refine_ls_number_restraints	0
_refine_ls_R_factor_all	0.0391
_refine_ls_R_factor_gt	0.0270
_refine_ls_wR_factor_ref	0.0792
_refine_ls_wR_factor_gt	0.0704
_refine_ls_goodness_of_fit_ref	1.066
_refine_ls_restrained_S_all	1.066
_refine_ls_shift/su_max	0.000
_refine_ls_shift/su_mean	0.000

Fig. 2.24 An edited extract of a Crystallographic Information File (CIF), giving experimental details for a crystal structure determination; the compound is [PhSNSNSNSPh][AsF$_6$], used at various points in this chapter as an example.

a complete CIF for a large structure is a long file, including not only the primary results, but also the derived geometry with associated s.u.s. The diffraction data can also be stored in a defined CIF format, as can an entire research report or manuscript for publication. Computer programs and journal publishing systems are then used for converting information from a CIF into text and tables more suitable for human readers, by-passing any need for manual typesetting with its inherent probability of introducing errors. Crystallography is particularly well suited to electronic publishing and has led the field in such developments.

The complete record of a crystal structure analysis encapsulated in a CIF also lends itself readily to a range of validation processes, contributing to the generally high reliability of X-ray crystallography as a structural tool in modern chemistry. Various computer programs and online facilities are available that will:

Many of these validation tests are combined in the online **CheckCIF** facility provided by the International Union of Crystallography.

- check a CIF for conformity to the agreed standards in terms of its contents, structure, and internal consistency;
- check for consistency of the unit cell geometry, space-group symmetry, and atom positions with the derived geometrical results (bond lengths etc.);
- check the geometry, displacement parameters, and other results against expected behaviour and typical ranges of values for such items, flagging unusual items for closer examination and possible correction.

Such validation procedures are a required part of the process of submitting structural results for publication in many journals, and help to avoid erroneous results appearing in the literature and in databases.

Once structural results have been published in the primary research literature (scientific journals), they are available for anyone to access and use. It is, however, not necessary to work through libraries of printed material to find results of relevance and interest, because of the availability of computer databases. Databases are essentially collections of items of information with a common structure and format. Their advantages over paper-based or other storage and retrieval systems include their ease of maintenance and updating, the possibility of automatic validation of new entries, facilities for selecting and sorting entries, and computer-based analysis of selected entries. A database has two components: the stored contents, and suitable software for search, retrieval, and analysis.

Computer databases are important in many areas of chemistry and other sciences, and cover such aspects as bibliography and literature citations, safety information, spectroscopy, and reaction mechanisms. They are particularly well suited for crystal structures.

Four main structural databases are used internationally in research. *CrystMet*, produced by Toth Information Systems Incorporated, holds information on metals, alloys, and intermetallics. In July 2014 there were over 150 000 entries. The *Inorganic Crystal Structure Database* (ICSD) is managed by the Fachinformationszentrum in Germany and the US National Institute of Standards and Technology. It contains inorganics and minerals, in which there is no organic carbon. In May 2014 there were over 160 000 entries. The *Protein Data Bank* (PDB), maintained by the Research Collaboratory for Structural Bioinformatics, stores data for proteins, nucleic acids, and larger biological assemblies; see Chapter 4 for a discussion of biological macromolecular crystallography. A relative newcomer, its size has grown enormously, with over 100 000 entries in August 2014. The largest structural database is the *Cambridge Structural Database* (CSD), developed by the Cambridge Crystallographic Data Centre, UK. Its contents are organics, organometallics, and metal complexes, and numbered over 700 000 in April 2014, with continued rapid expansion.

Each of these databases has its own individual special features appropriate to the contents, but they also have common aspects. Since the CSD is the most widely used, we consider it further here in illustrating some points. Entries for the CSD are drawn mainly from the primary research literature and are now almost entirely obtained in electronic form from the authors' submission. Other entries are supplied direct by crystallographers for inclusion in CSD as Private Communications, which remain otherwise unpublished. Individual entries are thoroughly checked for consistency and possible errors, which are either corrected or flagged.

Each entry in the database contains: bibliographic information; a collection of individual text and numeric data such as unit cell parameters, temperature of data collection, and *R* factor; a two-dimensional representation of the chemical structural formula; and all the atom positions in the asymmetric unit, from which detailed geometry can be calculated. Searches can be made through the contents against any of these items; particularly useful is the facility to search for all structures containing a particular group of atoms (a molecular fragment), with or without specific restrictions on its geometry. The possible output includes display of all the searchable items, a three-dimensional graphical representation of the structure which can be manipulated interactively, and statistical analysis of any of the numerical items, including specific geometrical features such as bond lengths.

There is, of course, much that the databases do *not* contain, such as the authors' discussion of their results, for which the original literature must be consulted; but even here, the databases provide the necessary bibliographic information as a way into the literature.

The structural databases are thus an invaluable resource of reliable information, far more convenient to use than the original published literature. They can be used to find a particular structure for various reasons (this includes avoiding repeating work which has already been done!), to obtain information on a series of related structures, to generate a geometry for a structural fragment for use in other calculations such as a Patterson search to solve a structure, molecular orbital theory, or molecular modelling, and for extensive research into trends and patterns in structures (such as conformations of rings, hydrogen bonding, intermolecular interactions, substituent effects, etc.).

Such research, based on extensive and often sophisticated analysis of structures in the databases, has uncovered numerous fascinating structural relationships and can even provide information relevant to chemical reactions through careful examination of the distortions produced in molecular structures by significant interactions with their crystal environment. It is often referred to as 'database mining'.

2.13 Summary

- The determination of a crystal structure by X-ray diffraction may be achieved within a day, or it may take much longer, depending on many factors including the quality of crystals, type of experimental equipment and X-ray source available, size and complexity of the structure, and problems encountered during structure solution, refinement, and interpretation.

- The sample must be a single crystal of appropriate size for the material and experimental setup; no other crystalline material should be present in the X-ray beam during data collection.

- Diffraction patterns are usually measured using an X-ray diffractometer, consisting essentially of an X-ray source, a device for rotating the crystal in the X-ray beam, an X-ray detector, and computer control. Most modern diffractometers have an electronic area detector, giving data collection times ranging from minutes to hours. Computer analysis of the diffraction pattern provides unit cell parameters, possible space groups, and a list of measured intensities with their associated reflection indices and standard uncertainties.

- Corrections may be required for variations in the incident X-ray intensity, crystal deterioration in the beam, various geometrical factors associated with the diffraction process, and other physical effects such as X-ray absorption. The corrected intensities are proportional to the squares of diffracted wave amplitudes, and experimental reflection phases cannot be measured.

- Structures may be solved (atom positions found) by a range of methods, of which the most common are analysis of the Patterson function for heavy atoms, so-called direct methods based on probability relationships among reflection phases, and dual-space methods exploiting the limited information available in both direct space (the crystal structure) and reciprocal space (the diffraction pattern).

- A structural model is refined by least-squares methods to give calculated amplitudes that match as closely as possible the observed amplitudes, the observations being appropriately weighted according to their perceived reliability. Refinement may incorporate constraints, restraints, and other aids to overcome difficulties such as the low scattering power of hydrogen atoms.

- Structure refinement is monitored by a range of statistical indicators known generally as R factors; the correctness of a final structural model is demonstrated by an essentially featureless difference electron density map.

- Problems often encountered in crystal structure determination, for which tools are available in refinement software, include static disorder and twinning.

- The absolute configuration of a chiral structure can be determined in X-ray crystallography by use of resonant scattering effects, which are significant and useful for certain combinations of chemical elements and X-ray wavelengths.

- The primary results of structure refinement are the positions (and displacement parameters) of independent atoms within a unit cell of known geometry and subject to the symmetry of a particular space group. From these, intramolecular and intermolecular geometrical parameters of interest can be calculated and analysed. Many different graphical representations of structures are possible, to suit the desired presentation of the results.

- The results of crystal structure determinations are conveniently recorded in a standard form known as the Crystallographic Information File (CIF), which serves as a vehicle for archiving, transmission, and publication. Published crystal structures are available in large internationally recognized computer databases, each with associated software for search, retrieval, display, and analysis.

- The overall effect is as if we could operate a microscope of around 10^8 magnifying power giving a result with a short delay rather than instantaneously.

2.14 **Exercises**

Exercise 2.1

Which steps in the flowchart of Fig. 2.1 have become generally faster as a result of improved computer hardware and software, and which have been largely unaffected by these developments?

Exercise 2.2

Why must a crystal be rotated in the X-ray beam during data collection (a) in order to obtain the complete diffraction pattern; (b) even in the measurement of a single reflection with a serial diffractometer?

Exercise 2.3

What advantages are there in measuring the complete diffraction pattern (with a full range of negative and positive values for all three indices, as would be necessary for a non-centrosymmetric triclinic structure in space group $P1$) rather than only the unique portion without symmetry-equivalent measurements?

Exercise 2.4

List the types of information that are available (a) in direct space and (b) in reciprocal space at the stage when X-ray diffraction data have been measured and 'reduced' (corrected) and the next step is to solve the structure.

Exercise 2.5

If possible, obtain a copy of the Excel spreadsheet described in Section 2.7 and a copy of the publication describing its use, and work through the stages outlined for the solution of the 1D structure.

Exercise 2.6

Why do Fig. 2.17(c) and (d) show only two of the six F atoms of the $[AsF_6]^-$ anion? Why are H atoms not found until all the other atoms have been located and refined?

Exercise 2.7

Give two reasons why crystallographic R factors never decrease to a value of zero, even with good quality data and the best available refinement software.

Exercise 2.8

Which of the following solvent molecules must be disordered, and which could be ordered?

- THF on a twofold rotation axis.
- THF on a mirror plane.
- *n*-pentane on an inversion centre.
- *n*-hexane on an inversion centre.

Exercise 2.9

Why are copper-target X-rays often used in preference to molybdenum-target X-rays for the study of natural products containing only C, H, N, O, and F atoms?

X-ray crystallography case studies

3.1 Introduction

In this chapter the process of X-ray crystal structure determination is illustrated by a series of five examples drawn from a wide range of structural chemistry research. The examples have been chosen to cover many different aspects of the experimental measurements and methods of structure solution and refinement, as well as a variety of types of material and some of the potential problems described earlier. Not all details of every structure determination are given, but each example presents particular features of interest. All the examples are of published work and references are provided as well as CSD REFCODEs so that further details can be found. 3D rotatable images and computer results and data files are also available in the supplementary electronic resources. At the end of the chapter (Section 3.8) some problems are presented for the reader to solve.

3.2 Case study 1: a mercury thiolate complex

The complex $[Et_4N][Hg(SR)_3]$ (Fig. 3.1), where R is the *cyclo*-hexyl group C_6H_{11}, is prepared from $HgCl_2$, NaSR and $[Et_4N]Cl$ in acetonitrile solution. This is an empirical formula corresponding to a monomeric structure; the anion could actually be a dimer with bridging thiolate ligands, a higher oligomer, or even a polymer, and this is one of the key questions to be answered by a crystal structure determination. Examination of a crystal of size $0.52 \times 0.36 \times 0.34$ mm on a four-circle serial diffractometer with molybdenum radiation of wavelength 0.71073 Å, at a temperature of 240 K, reveals a triclinic unit cell with dimensions

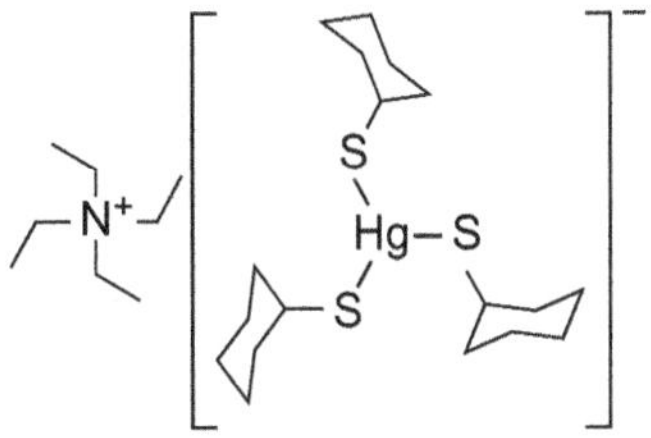

$$a = 10.724(4) \quad b = 12.440(5) \quad c = 12.643(5)\,\text{Å}$$
$$\alpha = 72.40(2) \quad \beta = 79.36(2) \quad \gamma = 73.33(2)°$$
$$V = 1531.3(10)\,\text{Å}^3$$

The formula mass for the proposed formula $C_{26}H_{53}HgNS_3$ is 676.5 daltons; this gives a calculated density of 1.467 g cm^{-3} and an average volume of 24.7 Å^3

Fig. 3.1 The expected chemical structure of case study 1.

The numbers in parentheses are standard uncertainties, expressed for compactness as units in the last figure of the corresponding numerical value. Thus, for example, 10.724(4) Å means a value of 10.724 Å with an s.u. of 0.004 Å.

In this space group symbol, the *P* means a primitive unit cell, and the 1̄ means an inversion centre as the only other symmetry. The other triclinic space group, *P*1, has only translation symmetry, no inversion; it is less likely because it would not require Z = 2 and because it is far less commonly found for non-chiral materials.

The subscript 'int' stands for *internal*, since this is a measure of the internal consistency of agreement of the data, not agreement with something else.

per non-hydrogen atom, if Z = 2; both of these values are reasonable for such a compound containing a heavy metal atom—the density was not actually measured. This means there are two cations and two anions in the unit cell. There are only two possible triclinic space groups (see Section 1.6), and the more likely is *P*1̄, which requires the two cations to be related to each other by inversion symmetry, and similarly for the two anions in the unit cell; the asymmetric unit of the structure (half the unit cell) is one cation and one anion, so we know nothing at this stage about the molecular geometry from symmetry arguments.

All possible reflections with θ < 25° have been measured one by one, including those equivalent by symmetry, a few of them more than once, giving a total of 10 990 reflections. Corrections are applied for absorption effects, which are strong for a compound containing mercury, based on measurements of intensities of selected medium-strong reflections at a range of different crystal orientations (these would be equal if there were no absorption, and an empirical correction can be calculated from their observed variation); the correction indicates that about 75–85% of the intensity of each reflection is lost by absorption of the incident and diffracted beams, the high values being due to the large absorption coefficient of the material and the range being a result of the crystal shape. It is also found that the intensities have decreased steadily by about 7% in total during the data collection period of around 4 days, and this is corrected for, so that all the data are on the same scale. Each pair of reflections with indices *h, k, l* and *−h, −k, −l* is equivalent by inversion symmetry, so they are averaged, to give a unique set of 5412 reflections. The averaging process also provides a measure of the agreement of symmetry-equivalent data in the form of a factor R_{int}, defined rather like the R factors in structure refinement, except that comparison is between pairs of observed symmetry-equivalent reflections instead of between observed and calculated values; the value 0.022 obtained for this set of data is excellent.

With just one heavy atom in the asymmetric unit, two in the unit cell, this structure is an obvious candidate for a Patterson synthesis as the means of solution. The largest peaks found in the Patterson synthesis (one half of the unit cell only; the other half is equivalent by inversion symmetry) are shown in Table 3.1.

Table 3.1 The largest Patterson peaks for case study 1

Peak no.	*u*	*v*	*w*	Peak height	Vector length (Å)
1	0.000	0.000	0.000	999	0.00
2	0.462	0.146	0.432	403	8.87
3	0.354	0.273	0.265	132	7.49
4	0.111	0.867	0.163	118	2.47
5	0.466	0.957	0.407	115	7.61
6	0.003	0.811	0.975	110	2.46
7	0.074	0.022	0.158	101	2.39
8	0.462	0.836	0.411	99	7.28

The peak heights are scaled arbitrarily so that the largest peak, at the origin, has a height of 999. The length given for each peak is the length of the corresponding interatomic vector, which is the distance (in Å) between the two atoms concerned.

In this space group, for each atom at a position x, y, z there is a symmetry-equivalent atom at position $-x, -y, -z$. The two mercury atoms in the unit cell thus have coordinates which are equal but opposite in sign, and the vector between these two positions is $x-(-x), y-(-y), z-(-z)$, which is just $2x, 2y, 2z$. The highest Patterson peak, therefore, excluding the origin peak, should correspond to an Hg–Hg vector, should be much larger than the other peaks, and has coordinates equal to twice those of a mercury atom in the structure; its vector length is the distance between the two mercury atoms in the unit cell. The coordinates of one mercury atom are thus 0.462/2, 0.146/2, 0.432/2, giving $x = 0.231$, $y = 0.073$, $z = 0.216$. The two mercury atoms are well separated (almost 9 Å apart); a short distance here would indicate that the two mercury atoms are in fact part of a dimeric anion of formula $[Hg_2(SR)_6]^{2-}$, probably with bridging thiolate ligands, so we can already deduce that the anion is monomeric, as proposed in Fig. 3.1.

A mercury atom has 80 electrons, and this is a significant proportion of the total scattering power of the asymmetric unit (344 electrons). Since we now know the position of the mercury atom, we can use this as our first model structure in the next stage of Fourier syntheses to find the remaining atoms. It is, however, worth pausing to examine the next highest peaks in the Patterson map. There are six of these with similar heights; all other peak heights are less than 60 on this scale. The next largest peaks after Hg–Hg are expected to be Hg–S. There should be three of these corresponding to *intramolecular* vectors, i.e. Hg–S bonds within one anion, and three corresponding to *intermolecular* vectors, i.e. from mercury in one anion to the three sulfur atoms in the other anion. The list does include three vectors of length 2.47, 2.46 and 2.39 Å, which are appropriate for bonds, and three vectors more than 7 Å long, which will be the intermolecular ones. So the Patterson map is certainly consistent with our proposed chemical formula and with a three-coordinate Hg-centred anion.

Now we move into the structure completion 'bootstrap' procedure. Taking the mercury atom alone as the model structure, Fourier transformation gives a calculated diffraction pattern. The value of the residual factor R1 is 0.284 for the 5032 reflections which have $F^2 > 2\sigma(F^2)$ (such reflections are sometimes called 'observed reflections', because they have an intensity judged to be significantly higher than background scattering), and wR2 is 0.650 for the complete set of reflections; these values are reasonable for a model structure containing only one heavy atom. A difference electron density map calculated from the observed amplitudes and the calculated phases derived from just the mercury atom (Fig. 3.2) does not show the complete electron density of the mercury atom (it would appear clearly as a very large peak in a full electron density map, but a difference map is better for finding new atoms). Its highest three peaks, with electron densities above 20 e Å^{-3}, are in positions about 2.4 Å from mercury, suitable for sulfur atoms; all 26 carbon atoms and the single nitrogen atom are among the 30 next highest peaks excluding residual electron density close to the position

It would be possible to use this information to calculate the positions of the three S atoms bonded to Hg and add them to the starting model structure, but it is not necessary.

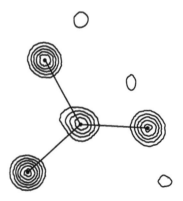

Fig. 3.2 Part of a 2D slice through the 3D difference electron density map around the mercury atom position; the points and lines show the final refined mercury and sulfur atom positions and bonds.

of mercury, with densities in the range 8.6 down to 2.8 e Å^{-3}, and the remaining peaks are rather lower. It is not expected that all carbon atoms will have the same maximum electron density, because those which undergo larger vibrations have their electron density spread out over a larger volume, so they usually show up as peaks of lower height. The assignment of atom types is made on the basis of the observed distances and angles involving the peak positions, and the expected structure, as well as on peak heights.

So in this case, a single cycle of Fourier synthesis calculations reveals all the non-hydrogen atoms, and refinement of the structure can begin. Refinement with isotropic displacement parameters for all the atoms (each atom has one adjustable overall displacement parameter as well as three adjustable coord-inates) reduces $wR2$ from 0.362 (with all atoms in the positions found from the difference map) to 0.311 and the value of $R1$ after refinement is 0.102, a consid-erable improvement on the first trial structure with just the mercury atom pres-ent. Inclusion of anisotropic displacement parameters (six values for each atom instead of one) reduces $wR2$ to 0.137 and $R1$ to 0.041; there are now 280 refined parameters (3 coordinates and 6 displacement parameters for each of 31 atoms, together with an overall scale factor to bring the observed and calculated intensi-ties onto a common scale).

Most of the 53 hydrogen atoms now show up in a further difference elec-tron density map, those in the anion being clearer than those in the cation, for which the atoms have higher displacement parameters, so their electron density is more spread out. They are all included in the refinement, but the C–H bond lengths and angles involving hydrogen atoms are kept fixed (con-strained) at ideal values rather than being allowed to refine freely, because the hydrogen atoms are not very precisely located by X-ray diffraction, especially in the presence of the heavy Hg atom; effectively, the hydrogen atoms are made to ride on their parent carbon atoms. This technique incorporates all the electron density of the atoms in the model structure, but it adds very few or no extra refined parameters; in this particular structure, free rotation is allowed about the C–C bonds of the cation starting from an idealized staggered confor-mation; such small deviations from ideal positions are due to intermolecular interactions. The final refinement also includes some extra minor corrections for effects which are not important for this account, including small modifi-cations to the relative weighting of different reflections, and gives values of 0.082 for $wR2$ (all data) and 0.032 for $R1$ (observed data). There are 285 refined parameters derived from the 5412 data, a very high degree of over-determi-nation, so the precision of the structure is high (s.u. values of the parameters are small). In a final difference electron density map there a few peaks of size 0.8–2.1 e Å^{-3} very close to the mercury atom (a commonly observed feature for heavy atom structures, due largely to an imperfect correction for strong absorption effects in the data), and no other peaks above 0.6 e Å^{-3}, which is an insignificant level.

The cation and anion are shown in Fig. 3.3. The main interest, in addition to the degree of association (a monomer rather than a dimer or higher aggregate) is in the coordination of the mercury atom, which is somewhat distorted trigonal

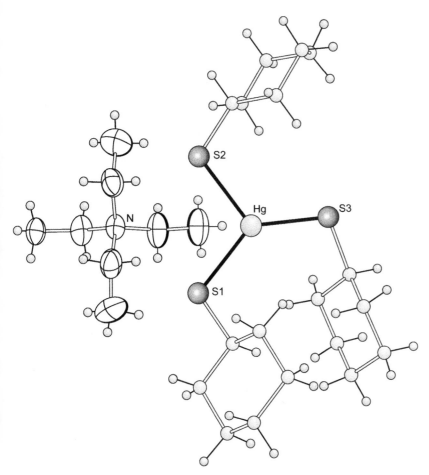

Fig. 3.3 The structure of the cation and anion of case study 1; the anion is shown in conventional ball-and-stick form, the cation with displacement ellipsoids (the displayed envelope enclosing 50% probability for each atom) for carbon and nitrogen atoms, to illustrate different styles of representation.

This structure has been published: Homoleptic cyclohexanethiolato complexes of mercury(II). T. Alsina, W. Clegg, K. A. Fraser and J. Sola, *J. Chem. Soc. Dalton Trans.* 1992, 1393–1399. The CSD REFCODE is VOXTOR.

planar, with a range of about 0.08 Å for the three Hg–S bond lengths, much greater than their individual uncertainties of about 0.002 Å, and three very different S–Hg–S angles, the smallest being 101.27(5)° and the largest 135.82(5)°. As expected, all the cyclohexane rings show a chair conformation with sulfur in an equatorial position. There is nothing at all unexpected or special about the geometry of the cation, which occurs in many salts of complex anions. It is an important feature of crystal structure determination that, in general, the whole structure has to be determined, even if only one particular part of it is really of interest; it is an all-or-nothing technique. This is a direct consequence of the nature of Fourier transforms: all the atoms in the unit cell contribute together to the observed diffraction pattern, and all the diffraction pattern has to be used to find the atoms.

In summary, this case study illustrates the following points:

- how simple it is to find a single heavy atom in the asymmetric unit of this common low-symmetry space group by a Patterson synthesis;
- the straightforward completion of the structure by Fourier methods from this one heavy atom as a starting point (phasing model);
- the necessity to locate and refine all the atoms of the structure even if some parts of it are of little or no particular interest;
- the importance of corrections for non-diffraction effects in the experiment.

3.3 Case study 2: a solvated chiral rhodium complex

This complex, with expected chemical formula $[Rh(Ph_2PCH_2PPh_2)(C_{12}H_8BO_4)]$ or $[Rh(dppm)(Bcat_2)]$, was prepared from $[(acac)Rh(dppm)]$ and $B_2(cat)_3$ in THF solution (see Fig. 3.4 for the ligands); crystals were obtained from a solution in C_6D_6, the deuterated solvent being used for NMR studies. The diphosphine dppm is a commonly used chelating ligand. The main interest in this structure is the coordination of Rh, especially the mode of attachment of the $Bcat_2^-$ ligand.

The crystal size was $0.36 \times 0.18 \times 0.16$ mm, and data were collected in a few hours with radiation of wavelength 0.71073 Å at 160 K, on a CCD-based area detector diffractometer. The total number of measured reflections is 26 554, giving 8415 unique data after the application of corrections for absorption (less severe than for case study 1, but still significant) and the averaging of symmetry-equivalent reflections, with $R_{int} = 0.025$.

The crystal system is orthorhombic and the space group is $P2_12_12_1$ (unambiguously indicated by the systematic absences in the data, with the lack of inversion

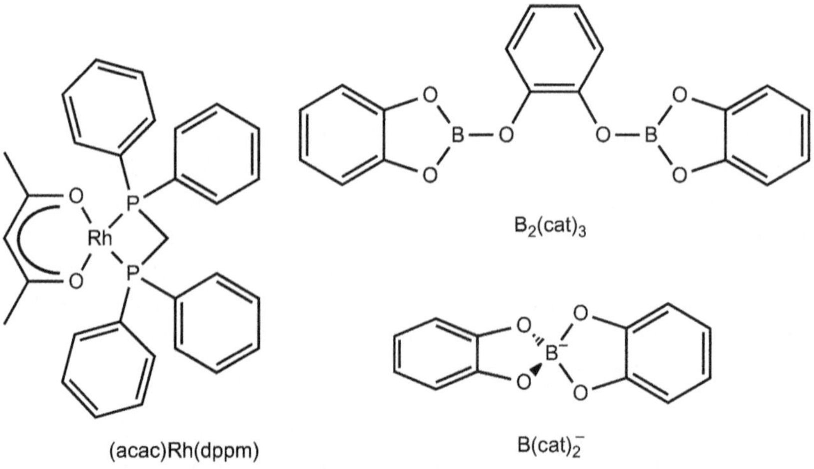

(acac)Rh(dppm) $B(cat)_2^-$

B$_2$(cat)$_3$

Fig. 3.4 Reagents and ligands for case study 2.

symmetry also indicated by the statistical analysis of intensities). The unit cell parameters are as follows:

$$a = 13.2932(7) \quad b = 15.2327(8) \quad c = 17.8046(10)\,\text{Å}$$
$$\alpha = \beta = \gamma = 90° \qquad\qquad V = 3605(3)\,\text{Å}^3$$

This cell volume is sufficient for 4 molecules of the proposed complex, with an average non-hydrogen atom volume of 20.1 Å^3, slightly on the high side for such a compound; the formula mass of 714.3 daltons gives a calculated density of 1.309 g cm^{-3}, which is reasonable. Reference to standard space group tables shows that the asymmetric unit is one-quarter of the unit cell, so we expect to find one molecule in the asymmetric unit, and there is no information at this stage about the molecular shape, all atoms lying in general positions (in fact there are no special positions in this space group, in which the only symmetry elements are screw axes and pure translation).

The presence of rhodium (45 electrons; the next largest atomic number is 15 for phosphorus) means a Patterson synthesis is again suitable for the solution of this structure (though, in fact, it can be solved easily by automatic direct methods or charge flipping to give most or all of the non-hydrogen atoms). For this space group there are four equivalent general positions in the unit cell, related by the screw axes in all three cell axis directions, with coordinates: x, y, z; $\frac{1}{2} - x, -y, \frac{1}{2} + z$; $-x, \frac{1}{2} + y, \frac{1}{2} - z$; $\frac{1}{2} + x, \frac{1}{2} - y, -z$. Four atom positions give 16 vectors as shown in Table 3.2, where each entry in the table body is the difference between the position at the top of the column and the position at the left of the row; wherever the number $-\frac{1}{2}$ would appear by this simple subtraction, it is replaced here by $\frac{1}{2}$ to give a neater table, because it is always permissible to add or subtract any whole number (in this case adding one) to a coordinate, which has the effect of moving to an exactly equivalent position in another unit cell.

Each row, and each column, contains one entry 0, 0, 0 for the vector between an atom and itself; one entry with $\frac{1}{2}$ as the first coordinate; one with $\frac{1}{2}$ as the second coordinate; and one with $\frac{1}{2}$ as the third coordinate. Some of the entries are identical except for a change of sign before one or more of 2x, 2y, and 2z. In fact, the 16 vectors give only 3 unique non-origin Patterson peaks together with their equivalents in orthorhombic symmetry; note that for every entry there is

Table 3.2. Vectors for equivalent atoms in space group $P2_12_12_1$

	x,y,z	$\frac{1}{2}-x,-y,\frac{1}{2}+z$	$-x,\frac{1}{2}+y,\frac{1}{2}-z$	$\frac{1}{2}+x,\frac{1}{2}-y,-z$
x,y,z	$0,0,0$	$\frac{1}{2}-2x,-2y,\frac{1}{2}$	$-2x,\frac{1}{2},\frac{1}{2}-2z$	$\frac{1}{2},\frac{1}{2}-2y,-2z$
$\frac{1}{2}-x,-y,\frac{1}{2}+z$	$\frac{1}{2}+2x,2y,\frac{1}{2}$	$0,0,0$	$\frac{1}{2},\frac{1}{2}+2y,-2z$	$2x,\frac{1}{2},\frac{1}{2}-2z$
$-x,\frac{1}{2}+y,\frac{1}{2}-z$	$2x,\frac{1}{2},\frac{1}{2}+2z$	$\frac{1}{2},\frac{1}{2}-2y,2z$	$0,0,0$	$\frac{1}{2}+2x,-2y,\frac{1}{2}$
$\frac{1}{2}+x,\frac{1}{2}-y,-z$	$\frac{1}{2},\frac{1}{2}+2y,2z$	$-2x,\frac{1}{2},\frac{1}{2}+2z$	$\frac{1}{2}-2x,2y,\frac{1}{2}$	$0,0,0$

The advantage of constructing the complete vector table is that it shows when some entries are completely identical; in such cases, the two or more vectors involved are parallel, and they contribute to a single peak of combined height. It is important to recognize this when matching the expected vectors for the space group with the peak list obtained from the experimental Patterson map.

The symmetry of all Patterson maps in the orthorhombic crystal system is such that changing the sign of any one, two, or all three coordinates of a peak gives an equivalent peak; the asymmetric unit of the Patterson map (which is smaller than the asymmetric unit of the crystal structure if the latter is non-centrosymmetric) is one-eighth of the unit cell. The unique peak positions listed here have been chosen from among symmetry-equivalents in order to make the calculations easy for demonstration purposes.

another with all the signs changed, which appears in a different row and different column, because the Patterson synthesis is always centrosymmetric even when the crystal structure (as in this case) is not. The unique vectors are listed in any one single row or any one single column, and it is necessary to consider only one row or one column to interpret the Patterson map; here we will take the first column.

The highest unique peaks in the Patterson synthesis calculated from the diffraction pattern of this compound are listed in Table 3.3; each of these has others equivalent to it by symmetry. All other peaks are under 80 in height.

Table 3.3 The largest Patterson peaks for case study 2

Peak no.	u	v	w	Peak height	Vector length (Å)
1	0.000	0.000	0.000	999	0.00
2	0.500	0.788	0.800	173	8.21
3	0.094	0.500	0.300	169	9.38
4	0.594	0.288	0.500	158	11.29

Finding the position of the rhodium atom in the asymmetric unit from these peaks is a matter of identifying each of the peaks with a corresponding entry in column 1 of Table 3.2. Peak 2 has $u = \frac{1}{2}$ and so corresponds to the fourth entry: $\frac{1}{2}$, $\frac{1}{2} + 2y$, $2z$. From this we obtain $y = 0.144$ and $z = 0.400$. Similarly peaks 3 and 4 correspond to entries 3 and 2 respectively, and we obtain the following results:

$$\text{from peak 2:} \quad\quad\quad y = 0.144 \quad z = 0.400$$
$$\text{from peak 3:} \quad x = 0.047 \quad\quad\quad z = -0.100$$
$$\text{from peak 4:} \quad x = 0.047 \quad y = 0.144$$

Choosing a different column of the vector table, or one of the rows, would lead to a different set of three coordinates for the Rh atom, but it would be equivalent by symmetry or by taking a different valid unit cell origin for this space group.

For this particular space group (this is not always the case), each entry provides us with two coordinates, and we obtain two indications for each of x, y and z. The results for x and y agree, but there are two different values for z. The reason for this is inherent in the process for solving the Patterson synthesis and is easily explained. Note that peak 3, for example, at 0.094, 0.500, 0.300, appears in the same position (and at symmetry-related positions) in every unit cell of the Patterson map. There is thus also a peak at 0.094, 0.500, 1.300, from which the above calculation gives x = 0.047 (as before) and z = 0.400. Because a coordinate is obtained by dividing by 2, ½ can be added to or subtracted from the answer to give an equally valid result. This corresponds to choosing a different possible allowed unit cell origin (in most space groups, symmetry elements occur regularly spaced at intervals of one-half a lattice repeat, so possible unit cell origin choices lie at intervals of ½ along some or all of the unit cell axes).

We now have a first model structure consisting of a single Rh atom at 0.047, 0.144, 0.400 and can begin the Fourier bootstrap procedure to find the remaining atoms. The Rh atom alone is a smaller proportion of the total scattering power of the asymmetric unit than we had with the Hg atom in case study 1, so it is not surprising that our first $R1$ factor is somewhat higher in this case, at 0.336, with $wR2 = 0.709$. The phases calculated from the Rh atom are also not very close to the correct (unknown) phases, and so the resulting difference electron density does not clearly show all the missing non-hydrogen atoms; in fact, there are just two large peaks at sensible positions for the P atoms of the dppm ligand, and the lighter atoms are not well defined yet. It is usually counterproductive to include dubious atoms in the model structure, so the next model consists of just the Rh and two P atoms, a modest but significant improvement. This reduces $R1$ to 0.256 and $wR2$ to 0.616, and now all the C, O, and B atoms of the ligands are clearly revealed in a difference map as 42 of the 46 highest peaks and are readily assigned the correct atom types.

Refinement of all these atoms, with anisotropic displacement parameters, gives values of 0.080 for $R1$ and 0.255 for $wR2$ and a further difference map now contains 6 peaks with heights between 3.5 and 5.3 e Å^{-3}, all other peaks being under 1.4 e Å^{-3} in height. These form a regular planar hexagon and must be a molecule of benzene or, more correctly, perdeutero-benzene, the solvent from which the crystals were grown. This solvent molecule was not recognized earlier because its atoms have rather higher displacement parameters and hence lower electron density maxima, and because its presence was not expected. Addition of the 6 extra C atoms to the model structure, with further anisotropic refinement, reduces $R1$ to 0.032 and $wR2$ to 0.104. At this stage all 36 H atoms are revealed in a difference map.

In the final refinement the H atoms are included with riding-model constraints as for case study 1, giving a total of 460 refined parameters and 8415 unique data, a satisfactorily high data/parameter ratio. $R1$ is 0.021 and $wR2$ is 0.047; both of these are substantially reduced by inclusion of the H atoms. There are no difference electron density peaks above 0.28 e Å^{-3}.

The space group is non-centrosymmetric and the arrangement of ligands around the Rh atom is chiral, so the structure is not identical to its enantiomer, which can be generated by inverting the signs of all coordinates of all the atoms. Refinement of the inverted model structure gives significantly higher $R1$ (0.028) and $wR2$ (0.065), and the absolute structure parameter for the correct enantiomer has a value of $-0.012(7)$, insignificantly different from zero and with a very small s.u., indicating a very high confidence of the absolute chirality assignment; Rh and P atoms have significant resonant scattering effects.

The molecular structure is shown in Fig. 3.5. Rhodium, in oxidation state +1, is chelated by the diphosphine ligand, as expected, and is η^6-coordinated (in a half-sandwich fashion) by one benzene ring of the Bcat_2^- ligand, giving an overall neutral complex. The four-membered RhP_2C ring is essentially planar, as are both halves of the Bcat_2 ligand, which are perpendicular to each other. The asymmetric unit also contains one C_6D_6 molecule, so the compound is a **solvate**; the solvent molecules occupy spaces between the complex molecules and contribute to the overall packing, but there are no particularly strong intermolecular interactions and the solvent molecule shows rather higher atomic displacements than the molecule of the complex (which is one reason why it was not located in the initial structure solution). Standard uncertainties for Rh–P and Rh–C bond lengths are 0.0006 and 0.002–0.003 Å respectively, and those for bonds between lighter atoms are 0.003–0.005 Å and up

Six of these are actually D atoms, but H and D (like isotopes of other elements) are indistinguishable in X-ray diffraction because they have the same electron density and differ only in their nuclei. The only impact of having D rather than H atoms is in the molecular mass and crystal density; it does not affect the crystal structure itself, except that H and D atoms, with different masses, will have slightly different atomic displacements.

This structure has been published: Rhodium catalyzed diboration of unstrained internal alkenes and a new and general route to zwitterionic $[L_2Rh(\eta^6\text{-catBcat})](cat=1,2\text{-}O_2C_6H_4)$ complexes. C. Dai, E. G. Robins, A. J. Scott, W. Clegg, D. S. Yufit, J. A. K. Howard and T. B. Marder, *Chem. Commun.* 1998, 1983–1984. The CSD REFCODE is FAJJOP.

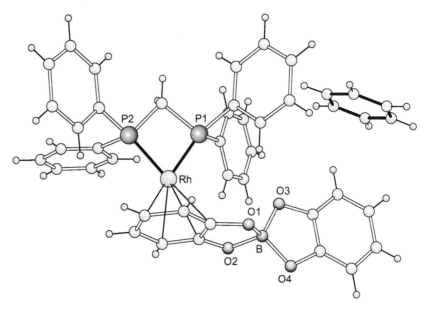

Fig. 3.5 The structure of the asymmetric unit of case study 2.

to 0.006 Å in the benzene solvent molecule; these values reflect the relative X-ray scattering factors of the different atoms and their atomic displacements.

In summary, this case study illustrates the following points:

- a typical Patterson solution for one heavy atom in a medium-symmetry space group, aided by analysis of vectors between symmetry-equivalent heavy atoms;

- routine completion of the structure by a few cycles of Fourier map calculations and addition of new atoms to the model structure;

- the incorporation of solvent molecules in a crystal structure to give a solvate, having a small effect on the crystal density and average atomic volume—in this case without significant intermolecular interactions;

- the equivalence of different isotopes of an element in X-ray diffraction;

- the determination of the 'absolute structure' of a chiral molecule by resonant scattering effects;

- the advantages of area detectors and low-temperature data collection in providing rapid and precise diffraction data.

3.4 Case study 3: a microcrystalline chiral organic compound

Tetracycline hydrochloride (Fig. 3.6) is an antibiotic agent. This is one of a number of compounds that were selected in 1998 for a competitive exercise in crystal structure determination from powder diffraction data collected by the organizers and made available to competitor research groups. It was important to have a definitive crystal structure obtained from single-crystal X-ray diffraction as a benchmark for

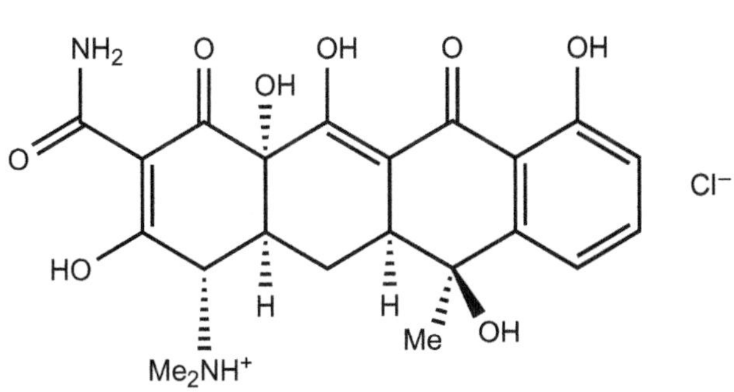

Fig. 3.6 The proposed chemical structure of tetracycline hydrochloride, case study 3.

the exercise, but at that stage no such information was available for the anhydrous salt. The sample provided was a commercial microcrystalline material, and recrystallization to generate larger single crystals was not permissible, as this could not be guaranteed to give exactly the same crystalline form. Since individual crystals were essentially coarse powder grains with maximum dimensions of tens of microns, conventional laboratory X-ray sources were unable to give adequate diffraction intensities, even though the crystal quality was good. Data were collected with synchrotron radiation ($\lambda = 0.6883$ Å) at 150 K, using a CCD-based area detector diffractometer of the same kind as for case study 2, from a crystal of dimensions $0.04 \times 0.03 \times 0.02$ mm.

The material is orthorhombic, with the same space group ($P2_12_12_1$) as for case study 2. The unit cell parameters are as follows.

$$a = 10.9300(9) \quad b = 12.7162(11) \quad c = 15.7085(13) \text{ Å}$$
$$\alpha = \beta = \gamma = 90° \qquad\qquad V = 2183.3(3) \text{ Å}^3$$

With a formula mass of 480.9 daltons, this gives a calculated density of 1.463 g cm^{-3} and an average volume of 16.6 Å^3 per non-hydrogen atom if $Z = 4$, corresponding to a single cation–anion pair in the asymmetric unit of the structure. The average atomic volume is somewhat lower than for a typical organic compound, and this is consistent with substantial hydrogen bonding in the structure, which reduces some of the intermolecular contact distances.

The total number of measured reflections is 8955, with 4915 unique data after averaging symmetry-equivalent reflections ($R_{int} = 0.047$). Corrections are not needed for absorption for such a small crystal and with no heavy atoms present, but they are required for a substantial steady decline in intensities, caused in this case not by sample decomposition in the X-ray beam but by a decay of the intensity of the X-ray source itself (a property of some older synchrotron facilities).

There are no particularly heavy atoms in this structure, chlorine having only about twice the electron density of the lighter non-hydrogen atoms. The structure is easily solved by direct methods: well-established programs available at the time this structure was originally determined require only the unit cell parameters, space group, X-ray wavelength, and an estimate of the contents of the asymmetric unit (or of the unit cell) together with the diffraction data in order to give a correct solution, while more recent direct methods and charge flipping programs are able to work out the space group as part of the process. A range of programs all reveal one large electron density peak for the chloride anion and 32 smaller peaks corresponding to the expected O, N, and C atoms; other peaks are much lower. This means the complete structure, other than H atoms, is found in a single calculation, and the Fourier bootstrap procedure is not needed.

Refinement proceeds as in the previous case studies; with anisotropic displacement parameters for all the non-H atoms, $R1 = 0.066$ and $wR2 = 0.173$. A difference map now shows 24 H atoms among the top 25 peaks; the position of one H atom is less clear, but it can be found in the next difference map after the other H atoms have been included in the model structure.

For this structure, the positions of H atoms bonded to O and N are of particular interest because of hydrogen bonding and because the compound has more than one possible **tautomeric form**, these being related by transfer of H atoms and exchange of single and double bonds. It is therefore important, if possible, to refine the H atoms without constraints or restraints, so that their positions are defined only by the experimental data and not by any preconceived ideas. This is successful here, with free refinement of the coordinates and isotropic displacement parameters of the 25 H atoms in a final calculation, which gives $R1 = 0.050$, $wR2 = 0.116$. The geometry involving H atoms, and their displacement parameters relative to their parent atoms, are all entirely reasonable; these H-atom parameters, of course, have rather higher standard uncertainties than those for the other atoms. The resonant scattering effects, although not very strong for Cl and almost non-existent for the lighter atoms, are sufficient to indicate the correct absolute configuration for this chiral compound, with an absolute structure parameter of 0.02(10), very close to zero and with an acceptably small s.u. Because of the free refinement of the H atoms, there are 399 refined parameters, but this still gives a high data/parameter ratio of 12.3.

The structure of the asymmetric unit (one cation, with protonation of the tetracycline molecule at the NMe_2 group, and one chloride anion) is shown in Fig. 3.7.

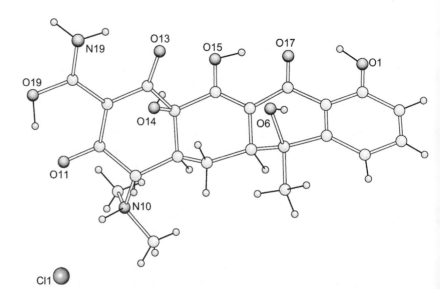

Fig. 3.7 The asymmetric unit of tetracycline hydrochloride, case study 3. N, O, and Cl atoms are labelled for comparison with the table of hydrogen bonds.

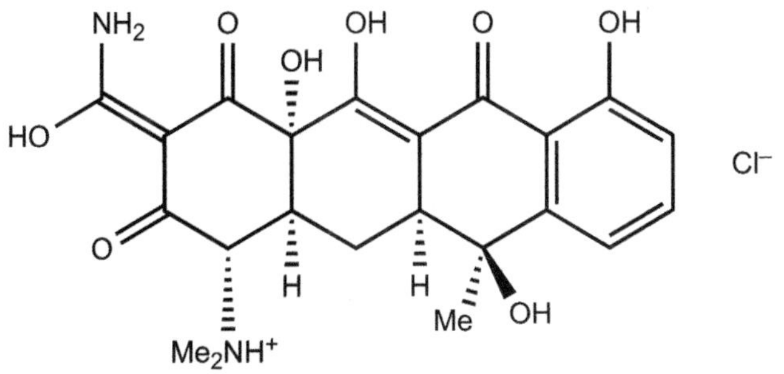

Fig. 3.8 The observed tautomeric form of case study 3.

The tautomeric form, based on the observed H atom positions and on C–C and C–O bond lengths, is found to be that shown in Fig. 3.8, which differs from the one originally proposed and found in other crystal forms of tetracycline and its salts. The difference consists in the transfer of one H atom from a ring OH substituent to the adjacent amide $CONH_2$ group, with interconversion of some single and double bonds. Interestingly, this H atom forms a relatively strong O–H...O hydrogen bond (as indicated by a long O–H bond, a short H...O contact, and a short O...O distance) to the O atom from which it has been transferred, so the transfer may be regarded as incomplete; this, not surprisingly, was the last H atom to be found in the structure determination and it has the highest displacement parameter of all the atoms, corresponding to a relatively shallow potential energy minimum for its position between the two O atoms.

The hydrogen bonding in a section of the crystal structure of case study 3 is shown in Fig. 3.9; apart from the involvement of the chloride anion, all the hydrogen bonding is intramolecular. The geometry of hydrogen bonding is usually characterized by the X–H, H...Y and X...Y distances and the X–H...Y angle, where X is the hydrogen bond donor atom (here N or O) and Y is the acceptor (here O or Cl); this information is provided in Table 3.4 for tetracycline hydrochloride. In providing this information for intermolecular hydrogen bonds (and for intermolecular interactions generally), it is necessary to specify any symmetry operations relating the acceptor to an equivalent atom in the 'home' asymmetric unit.

In summary, this case study illustrates the following points:

- the use of intense synchrotron radiation for investigation of small crystals and other weakly scattering samples;
- a typical straightforward direct methods solution of an 'equal-atom' structure;
- the determination of absolute configuration from resonant scattering effects;
- the free refinement of H atoms in appropriate cases;
- the study of intramolecular and intermolecular hydrogen bonding in a crystal structure;
- the existence of related crystal forms such as polymorphs, hydrates and other solvates, and co-crystals.

There are several other published crystal structures that contain either tetracycline itself or its protonated cation along with an anion in the latter case and with solvent or other neutral molecules; 20 are found in the CSD, including a hexahydrate of the neutral molecule, a hydrated hydrochloride, and a whole series of **co-crystals** containing carboxylic acids. The study of **polymorphs,** solvates, co-crystals and other closely related crystal forms is an important pursuit with respect to variations in physical properties, especially in the pharmaceutical industry.

This structure has been published: Tetracycline hydrochloride: a synchrotron microcrystal study. W. Clegg and S. J. Teat, *Acta Crystallogr. Sect. C* 2000, **56**, 1343–1345. The CSD REFCODE is XAYCAB.

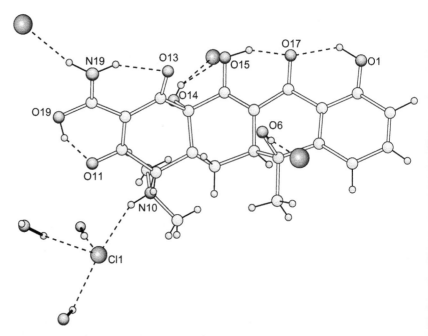

Fig. 3.9 Intramolecular and intermolecular hydrogen bonding (shown as dashed lines) in the crystal structure of case study 3. For hydrogen bonds to chloride formed by OH and NH_2 groups of symmetry-related cations, only these donor groups of atoms are shown; inclusion of complete cations would cause major congestion of the diagram.

Table 3.4 Hydrogen bonds for tetracycline (Å and °).

	X–H	H...Y	X...Y	X–H...Y
O1–H1...O17	0.92(6)	1.79(6)	2.564(5)	140(5)
O6–H6...Cl1a	0.94(6)	2.24(6)	3.179(3)	174(5)
O14–H14...Cl1b	0.70(5)	2.55(5)	3.195(3)	153(6)
O14–H14...O15	0.70(5)	2.36(6)	2.683(4)	110(5)
O15–H15...O17	0.95(6)	1.62(6)	2.491(4)	151(5)
O19–H19...O11	1.13(7)	1.49(7)	2.484(4)	143(5)
N10–H10N...Cl1	0.87(5)	2.26(5)	3.053(4)	152(5)
N19–H19A...Cl1c	0.92(6)	2.27(6)	3.157(4)	163(5)
N19–H19B...O13	0.87(6)	1.97(6)	2.694(5)	141(5)

Symmetry operations for equivalent atoms: a $x-1, y, z$; b $-x + 3/2, -y + 1, z + 1/2$; c $x-1/2, -y + 3/2, -z$

3.5 Case study 4: a metal coordination chain polymer

There is much current interest in the structures and properties of coordination polymers, especially where these contain channels or pores that could potentially be used for storage of gas molecules, with energy (hydrogen storage) and environmental (CO_2 storage) applications. Such compounds are generated when metal centres (nodes) are connected by multidentate ligands (linkers) that serve as bridges across two or more metal ions rather than chelates to a single metal ion. Depending on the coordination geometry and the nature and geometry of the linker ligands (and any non-bridging ligands that may also be present), the polymers may be one-, two- or three-dimensional; the higher-dimensional structures are also known as metal organic frameworks (MOFs). The polymeric structures may be electrically neutral, or they may carry a net charge balanced by uncoordinated counter-ions, which can occupy voids in the network along with solvent or other molecules.

The compound $[(MoS_4Cu_3I)_2(bbd)_3]$ uses a tetranuclear metal cluster rather than a single metal ion as a node, and the bridging ligand bbd shown in Fig. 3.10. It was prepared from $(NH_4)_2[MoS_4]$ and CuI (to generate the cluster node) together with bbd in dimethylformamide (DMF, Me_2NCHO) solution, and obtained as very small crystals requiring synchrotron radiation for data collection, carried out at 150 K.

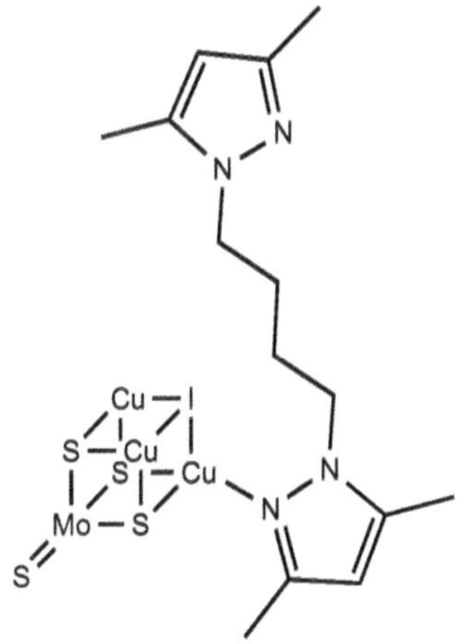

The crystals are monoclinic, space group $I2/a$, with the following unit cell parameters.

$$a = 19.937(6) \quad b = 10.001(3) \quad c = 35.707(14)\,Å$$
$$\beta = 97.164(4)° \quad \alpha = \gamma = 90° \quad V = 7064(4)\,Å^3$$

Chemical analysis results fit reasonably well with the formula given above, though slightly better if 1 mol of DMF solvent is included. With and without the added DMF, and for $Z = 4$, the calculated density and average non-hydrogen atomic volume are 1.783 and 1.714 g cm^{-3}, and 22.9 and 24.5 Å^3, respectively, which are reasonable for such a compound containing several large metal and iodine atoms. If the correct space group is indeed $I2/a$, then the expected value of Z is 8 for atoms in general positions, and so the two clusters in the chemical formula must be symmetry-equivalent, as are two of the three bbd ligands, with the third one lying on either an inversion centre or a twofold rotation axis; the asymmetric unit contains one cluster and 1.5 bbd ligands. The consequences of the space group symmetry for any DMF solvent present are discussed later.

Corrections were made for absorption (the crystal was a thin plate, $0.08 \times 0.08 \times <0.01$ mm) and other factors in the synchrotron data collection. The total number of measured reflections is 31 049, from a data collection taking less than one hour.

The structure is readily solved (and the space group thus confirmed) by standard methods. However, not all the atoms appear clearly in electron density maps: the $(CH_2)_4$ linker chain of the bbd ligand lying in a special position across an inversion centre (such that the two halves of the ligand are symmetry-equivalent) is not easily found at first; it has two independent C atoms, one of which is attached to the pyrazole ring and presents no great problem, but the other is found only after the rest of the atoms have been refined, and it has a low electron density maximum and a high refined displacement parameter. Isotropic refinement of all these atoms gives an $R1$ factor of 0.172, with $wR2 = 0.454$; introduction of anisotropic displacement parameters reduces these to 0.068 and 0.240, respectively. At this stage it is clear from the elongated shape of displacement ellipsoids

Fig. 3.10 The expected metal cluster node and bbd ligand of case study 4.

Space group $I2/a$ has a body-centred (I) unit cell, glide planes perpendicular to the b axis, and both simple twofold and screw axes parallel to this axis; it also has inversion centres. A non-centrosymmetric space group Ia shows the same systematic absences, but is less likely on the basis of a statistical analysis of the measured intensities, which supports the presence of inversion symmetry. The asymmetric unit for $I2/a$ is one-eighth of the unit cell. An alternative choice of axes can be made that gives the more conventional setting $C2/c$ for this space group, but this involves a larger value for the angle β.

that the centrosymmetric bbd ligand is disordered. It is possible to split each of the C and N atoms into two alternative positions, maintaining a sensible and similar geometry for the combinations of these into two disorder components of the ligand. Subsequent refinement with appropriate restraints on the geometry and on the relative sizes and shapes of displacement ellipsoids of neighbouring atoms gives a small reduction in R1 to 0.067 and wR2 to 0.235, but the important point is that all the atoms now have reasonable displacement parameters and the C_4 chain has a sensible geometry, so this is a better structural model. The two disorder components do not have equal occupancies, but are in a ratio of about 3:2.

The R factors are still a little high, and there are some difference electron density peaks of significant size lying too far away from any of the atoms in the model structure to be bonded to them. These are probably due to solvent molecules, but the collection of peaks does not resemble the geometry of a DMF molecule at all, so the solvent must be highly disordered. Before dealing with this, it is necessary to make the model structure as complete as possible, and this involves adding the hydrogen atoms. These are inserted in geometrically calculated positions and refined with a riding model as in case studies 1 and 2, allowing ligand methyl groups to rotate about the C–C bonds joining them to the pyrazole rings. R1 is now 0.064 and wR2 is 0.218.

The main contributor to the high R factors is probably unmodelled solvent. There is a group of difference electron density peaks in a suitable region of the structure not occupied by the atoms in the model, two at 3–4 and several at around 1 e $Å^{-3}$, but they make no geometrical sense. In such a case we turn to the procedure described briefly in Chapter 2, in which a Fourier transform is carried out on this region of electron density considered to be occupied by disordered solvent, and this is used as a contribution to the calculated diffraction pattern along with the normal Fourier transform of the model structure. For this structure, the very satisfactory result is a final R1 factor of 0.039, with wR2 = 0.103; the highest residual peaks in the final difference map are around 1 e $Å^{-3}$ and lie close to the heavy atoms (I and Mo), as is often the case. The disordered solvent fitting procedure indicates that there are four symmetry-equivalent solvent-accessible voids in the model structure, each lying on a twofold rotation axis, having a volume of about 300 $Å^3$, and containing approximately 50 electrons; these figures are an acceptable estimate of the requirements of one DMF molecule disordered in each of these voids, in agreement with the chemical analysis. As a DMF molecule has no twofold rotation axis, it must be disordered in such a position, and in this case the disorder cannot be resolved as two rotation-related overlapping components by inspection of the difference map obtained earlier. The chemical formula can thus be given with some confidence as $[(MoS_4Cu_3I)_2(bbd)_3]\cdot DMF$, with Z = 4.

The structure is best described as being composed of cube-like clusters connected by bridging bbd ligands to form a polymeric chain; the connections are made by Cu–N bonds, all three Cu atoms of each cluster being coordinated by pyrazine N atoms (Fig. 3.11). Pairs of clusters are connected by pairs of bbd ligands to form a $(cluster)_2(bbd)_2$ unit with twofold rotation symmetry,

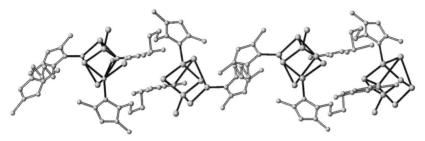

Fig. 3.11 Part of the polymeric chain structure of case study 4, showing only the major component of the disordered bbd ligand; H atoms are omitted.

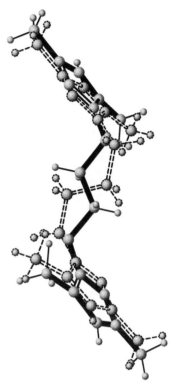

and these units are linked together by single bbd ligands, each of which is disordered over two different conformations across an inversion centre (in about a 57:43% ratio). The main feature of the bbd disorder is in the positions of the two central C atoms of the $(CH_2)_4$ linker chain between the pyrazine rings. This may most easily be understood in terms of the arrangements commonly found for torsion angles of saturated alkyl chains: adjacent C atoms usually have a staggered arrangement, such that X–C–C–X torsion angles are close to $\pm 60°$ (gauche, *g*) or 180° (anti, *a*). The sequence of three torsion angles in the NCCCCN chain of the ordered bbd ligand in this structure is *aga*, while the two components of the disordered bbd ligand have *aaa* and *gag* sequences (Fig. 3.12), giving three different conformations for this ligand within the structure. The cluster is of a well-known type, in which S atoms occupy three of the eight vertices of a distorted cube, three are occupied by Cu, one by Mo, and one (opposite Mo) by I.

Two other points of interest regarding this structure arise, not from the experiment or the structure itself, but from comparison with other structures. First, this is one of two polymorphs that have been found; in each of them the coordination polymer chain takes essentially the same form, with some minor differences in the details of the conformation, but the chains are packed together in different ways, giving a different unit cell (with one axis approximately halved and another approximately doubled in length) and a different space group for the second polymorph, in which the DMF solvent is ordered in a general position (not on a crystallographic symmetry element). Second, the corresponding tungsten compound (with W instead of Mo) adopts the same two polymorphic structures with very similar cell parameters, the same pair of space groups, and atoms in essentially the same positions. For each pair of polymorphs, the two compounds (containing Mo and W) are said to be **isostructural**. Pairs, and indeed whole sets, of compounds with chemical formulae differing only in the identity of one or more of the elements present are frequently isostructural; another example is the compound $[Cr(NH_3)_6][HgCl_5]$, used as an example in Chapter 1 for considerations of unit cell geometry and symmetry arguments, the cubic polymorph of which is isostructural with the corresponding compounds in which Cd or Cu replaces Hg in the anion, but not with the Zn compound nor with the compound in which Co replaces Cr in the cation (there is also a monoclinic polymorph).

Fig. 3.12 The two disorder components of the centrosymmetric bbd ligand. The minor component is shown with dashed circles for atoms and dashed bonds.

If two crystal structures have very similar unit cell parameters and the same space group, they are **isomorphous**; if, in addition, the atoms lie in essentially the same positions in the two structures, so that each of them can be used as a starting model structure (with appropriate changes in scattering factors for the substituted atoms) for the refinement of the other, they are isostructural. Note that the word **isostructural** may be used with different meanings in subjects other than crystallography.

The four crystal structures (two polymorphs each, for compounds containing Mo and W) have not all been determined in full. The existence of two pairs of polymorphs, and the isostructural relationships, have been deduced from a combination of two single-crystal structure determinations and some X-ray powder diffraction studies, a related technique described in Chapter 4.

This structure has been published: Metal-to-ligand ratio as a design factor in the one-pot synthesis of coordination polymers with $[MS_4Cu_n]$ (M = W or Mo, n = 3 or 5) cluster nodes and a flexible pyrazole-based bridging ligand. A. Beheshti, W. Clegg, V. Nobakht and R. W. Harrington, *Cryst. Growth Des.* 2013, **13**, 1023–1032. The CSD REFCODE is XIBVUB. Some of the computer programs and their combinations used here were not available at the time of the original research, so these results are not identical to those published (they are actually an improvement, particularly in the disorder modelling).

In summary, this case study illustrates the following points:

- an example in which parts of the structure lie in special positions on symmetry elements, with $Z' < 1$ (in this case 0.5);

- occurrence of disorder that can be modelled by the use of atoms with partial occupancy in alternative positions;

- the use of restraints and constraints to assist the refinement of a disordered structure;

- extensive disorder that cannot be modelled with discrete atom positions, requiring the use of a Fourier transform to calculate the contribution of the disordered region to the diffraction pattern;

- a polymeric structure, in this case one-dimensional;

- isostructural and polymorphic compounds.

3.6 Case study 5: a palladium complex of a bulky phosphine for catalysis studies

This complex, with a chemical formula $[PdCl(C_{12}H_{10}N)(C_{28}H_{35}P)]$ (Fig. 3.13) and synthesized in a mixed acetone-dichloromethane solvent, is one of a series of palladium complexes of interest for the efficient catalysis of organic cross-coupling reactions. It is another example of a sample obtained only as very small crystals ($0.05 \times 0.02 \times 0.02$ mm for the crystal examined) and thus requiring data collection with synchrotron radiation. At 120 K, the crystals are triclinic with the following unit dimensions.

$$a = 10.0924(18) \quad b = 16.907(3) \quad c = 20.136(4) \text{ Å}$$
$$\alpha = 88.004(2) \quad \beta = 89.567(2) \quad \gamma = 87.631(2)°$$
$$V = 3430.8(11) \text{ Å}^3$$

Simple calculation, as in previous examples, demonstrates that $Z = 4$, giving a density of 1.380 g cm^{-3} and an average non-hydrogen atomic volume of 19.5 Å^3. The two possible triclinic space groups have 1 (for $P1$) and 2 (for $P\bar{1}$) asymmetric units per unit cell, so in either case we have here a structure with $Z' > 1$ (4 or 2, respectively).

This case study presents a new problem: the reflections observed in the diffraction pattern cannot all be indexed on the basis of any reasonable single unit cell. The unit cell given above accounts for some of the reflections when it lies in one particular orientation (the indexing procedure determines simultaneously the unit cell parameters and the orientation), and others when it is rotated by 180° into a different orientation. Twofold rotation symmetry does not occur in a triclinic lattice, and so this is a twinned crystal with two twin components related in this way. The twin law is a 3×3 rotation matrix expressing the 180° rotation and its axis direction, and an initial estimate of the twin fraction can be found by comparing the average intensities of the two sets of non-overlapping reflections, which in

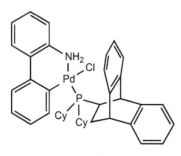

Fig. 3.13 The molecule of case study 5.

this case are approximately equal. In the data file of measured intensities derived from the data collection and reduction, individual reflections are labelled to show whether they belong to the first or second twin component, or have contributions from both components overlapping, with two sets of reflection indices assigned.

In the large collection of measured reflections, over 17 000 belong exclusively to the first component, essentially the same number to the second, and almost as many are overlaps (these numbers include symmetry equivalents and reflections measured more than once in different combinations of crystal and detector positions). Corrections are made in the usual way for absorption and other effects, and there are about 51 000 reflections in the complete set of data, 25 242 being unique (with h, k, l and $-h, -k, -l$ equivalent by symmetry in the space group $P\bar{1}$, subsequently confirmed as correct; R_{int} is 0.048 for the merging of equivalent reflections).

Once the twinning is recognized and appropriately treated, structure solution and refinement proceeds without difficulty; the structure can be solved by Patterson or direct methods, using data from one of the twin components, and refinement requires no restraints or constraints other than a standard riding-model treatment of hydrogen atoms. There is no disorder. Intermediate steps do not need to be described here.

The twin fraction is one of the refined parameters, and its final value is 0.4523(6), so we have a 55:45% two-component twin. Apart from recording this fact, it has no impact on the quality and interpretation of the structural results. The final values of $R1$ and $wR2$ are 0.052 and 0.139, respectively. The largest residual difference electron density peaks lie close to Pd atoms and near a cyclohexyl ring that could be subject to minor unresolved disorder, a possibility indicated also by somewhat elongated displacement ellipsoids, but validation of the structure does not flag this as a significant problem worth investigating further. These results are excellent for a challenging sample.

The asymmetric unit contains two chemically identical but crystallographically independent molecules, shown in their observed relative positions in Fig. 3.14. A least-squares fit of the two molecules, with one of them rotated and translated to overlay the other as closely as possible, is shown in Fig. 3.15, and demonstrates

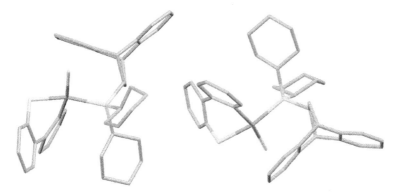

Fig. 3.14 The two molecules in the asymmetric unit of case study 5. H atoms are omitted.

This structure has been published: Electron-rich trialkyl-type dihydro-KITPHOS monophosphines: efficient ligands for palladium-catalyzed Suzuki–Miyaura cross-coupling. Comparison with their biaryl-like KITPHOS monophosphine counterparts. S. Doherty, J. G. Knight, N. A. B. Ward, D. M. Bittner, C. Wills, W. McFarlane, W. Clegg and R. W. Harrington, *Organometallics* 2013, **32**, 1773–1788. The CSD REFCODE is HOQQOV.

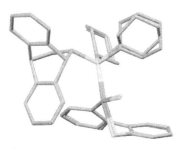

Fig. 3.15 A least-squares overlay of the two independent molecules, showing their very similar geometry. H atoms are omitted.

that they have almost exactly the same geometry, the main difference being in the orientation of one of the cyclohexyl rings. In structures with $Z' > 1$ this is by no means always the case, and conformational differences in molecules within the same crystal structure, as well as in different crystal structures, can be very interesting; there are even cases in which different isomers are found in the same structure.

In summary, this case study illustrates the following points:

- extraction of a correct unit cell in two orientations for a twinned crystal, thereby establishing the twin law;
- use of the complete two-component twinned data in refinement of the structure, including determining the twin ratio;
- the presence of two crystallographically independent molecules in the asymmetric unit ($Z' = 2$) and comparison of their structures.

3.7 Summary

This chapter has described the main features of crystal structure determinations of five example compounds, to illustrate the various stages described in Chapter 2 and some further points of interest. The following topics have been covered:

- calculations of unit cell contents and deductions about the presence of solvent and possible symmetry restrictions on molecular structure;
- some aspects of space group symmetry;
- the measurement of diffraction patterns with both serial and area-detector diffractometers;
- the use of different sources of X-rays;
- correction of measured intensity for effects such as absorption;
- the solution of crystal structures by Patterson, conventional direct, and dual-space methods;
- the completion of partial structure models with Fourier calculations;
- crystal structure refinement with isotropic and with anisotropic displacement parameters;
- the inclusion of hydrogen atoms, with and without constraints depending on circumstances;
- indicators of precision and completeness in structural results;
- a range of typical features found in crystal structures;
- the use of resonant scattering effects to determine absolute configuration and related properties;
- solvates;
- **polymorphism**;
- isostructural compounds;

- disorder and twinning as problems to be recognized and overcome;
- structures in which the asymmetric unit contains only part of a symmetric molecule, or contains more than one molecule of the same compound;

3.8 Exercises

Exercise 3.1

For case study 1, calculate the relative heights expected for non-overlapping Patterson peaks due to the following pairs of atoms: Hg–Hg, Hg–S, Hg–N, S–S. Confirm that the peaks listed in Table 3.1 have appropriate heights for Hg–Hg and Hg–S vectors. What should be the approximate height of the peaks appearing next in the list, and what atom pairs are responsible for them?

Exercise 3.2

The Hg–S bond lengths in case study 1 are obtained, at the end of the refinement, with higher precision (lower s.u. values) than the N–C and C–C bond lengths. Why is this?

Exercise 3.3

Why is it reasonable to expect that the $[Hg(SR)_3]^-$ anion of case study 1 will deviate significantly from showing perfect threefold rotation symmetry?

Exercise 3.4

The triclinic space group for the complex $[(C_{18}H_{18}N_4S)HgBr_2]$ in Exercise 1.4 is $P\bar{1}$. For each mercury atom at a position (x, y, z) in the unit cell, space group symmetry requires that there is another mercury atom at the position $(-x, -y, -z)$. Where, apart from the origin $(0, 0, 0)$, will the largest peaks be found in the Patterson map for this structure? The largest peaks found in the Patterson map calculated from the observed diffraction pattern are listed below (Table 3.5); there are also peaks at symmetry-equivalent positions. All other peaks are under 100 in height. Deduce the coordinates of one mercury atom in the structure. To what are peaks 3 and 4, and peaks 5 and 6, probably due?

Table 3.5 The largest Patterson peaks for Exercise 3.4

Peak number	x	y	z	Peak height	Vector length (Å)
1	0.000	0.000	0.000	999	0.00
2	0.358	0.374	0.540	336	8.23
3	0.118	-0.124	0.154	224	2.61
4	0.188	0.111	-0.094	223	2.50
5	0.452	0.514	0.558	223	8.46
6	0.471	0.243	0.689	222	6.76

Exercise 3.5

The monoclinic space group for the indium complex in Exercise 1.5 is $P2_1/c$. For each atom at a general position (x, y, z) in this space group, there must be three symmetry-equivalent atoms at positions $(-x, -y, -z)$, $(-x, ½ + y, ½ - z)$, and $(x, ½ - y, ½ + z)$. Derive from these the positions of the corresponding Patterson vector peaks (similar to Table 3.2, but with different entries). The four largest peaks in the Patterson map for this compound are at positions given in Table 3.6, together with peaks at symmetry-equivalent positions. Propose (x, y, z) coordinates for one indium atom consistent with these peaks.

Table 3.6 The largest Patterson peaks for Exercise 3.5

Peak number	x	y	z	Peak height	Vector length (Å)
1	0.000	0.000	0.000	999	0.00
2	0.000	0.888	0.500	348	8.45
3	-0.120	0.500	0.820	329	9.33
4	-0.120	0.388	0.320	179	8.77

Exercise 3.6

What difference is there in the following geometrical parameters for two enantiomers?

- Bond lengths.
- Bond angles.
- Torsion angles.

Exercise 3.7

Distinguish between isomers and polymorphs. Why is it important to investigate polymorphism and the formation of solvates for pharmaceutical compounds?

Exercise 3.8

What do you think might be the consequences of overlooking twinning in a diffraction pattern, and measuring and using only the reflections that fit the unit cell and orientation of one of the twin components?

4 Related topics

4.1 Introduction

The previous chapters have described the main topics of X-ray crystallography of interest to undergraduate chemistry students, under the headings of fundamentals, practical steps involved, and selected examples as case studies. In this final chapter, four further topics are considered briefly for the sake of completeness, extending the basic treatment beyond the core subject of X-ray single-crystal diffraction for the determination of chemical compounds. These are: the use of neutrons instead of X-rays (extending the choice of radiation source); the use of powder diffraction (extending the scope regarding the physical state of the sample); applications to biological macromolecules (extending the scope of the technique beyond what is sometimes called 'small-molecule' crystallography); and crystal structure prediction (extending the investigations to add theoretical to experimental methods).

4.2 Single-crystal neutron diffraction

X-rays are used for crystal structure determination because they have wavelengths comparable to the separations between atoms in molecules, and so they give measurable diffraction effects from crystals. Any other radiation with a similar wavelength would, in principle, serve the same purpose. Of course, there are no other forms of electromagnetic radiation with the same wavelengths as X-rays, by definition.

According to the **de Broglie relationship**

$$\lambda = h / p = h / mv \tag{4.1}$$

an object of mass m moving with velocity v and momentum $p = mv$ has an associated wavelength and can display corresponding wave properties. For neutrons generated by a nuclear reactor or a **neutron spallation source**, the associated wavelengths lie in the same range as X-rays, so a beam of neutrons can be diffracted as **particulate radiation** by crystalline material.

The use of neutrons for diffraction is experimentally much more difficult and expensive than the use of a conventional laboratory X-ray tube or even

Neutrons (and other elementary particles) from nuclear reactors have been used for scattering, diffraction and spectroscopy in research for many years; the reactor may be designed specifically for this purpose, or may be used primarily for energy generation or nuclear reactions. More recently neutron spallation sources have been developed, in which particles (typically protons) are accelerated in a synchrotron and fired in pulses at a target to generate neutrons and other useful particles. The pulsed nature of these neutron beams is a property exploited in specialized experiments; it provides, among other uses, a means of measuring neutron wavelengths by their de Broglie relationship to velocity through time-of-flight measurements.

a synchrotron source and, in most cases, diffracted intensities are considerably weaker, so there is no point in it unless it offers some significant advantages over X-ray diffraction. For most structure determinations this is not the case, and X-ray diffraction is much more widely used. There are, however, circumstances in which neutrons provide clear advantages, arising from the different ways in which neutrons and X-rays interact with matter as they pass through it.

X-rays, as we have seen, are scattered by the electrons of atoms; an X-ray diffraction experiment shows the electron density distribution within the unit cell of a crystal. This electron density distribution is usually interpreted in terms of atomic positions, and leads to molecular geometry. Since the electron density of each atom is generally distributed approximately symmetrically about the nucleus, this interpretation is valid, but in reality there are deviations from spherical symmetry, caused by chemical bonding and other valence effects. The effect is particularly marked for hydrogen atoms, which are consistently located too close to their bonded atoms by X-ray diffraction (Fig. 4.1).

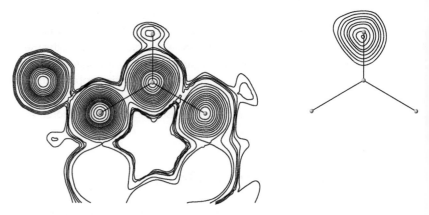

Fig. 4.1 Total electron density (left) and difference electron density (right) for the location of a hydrogen atom attached to a benzene ring, as obtained from X-ray diffraction at low temperature. The points and lines show the final refined positions of the atoms and bonds, with the C–H bond length extended to its expected internuclear distance (from spectroscopic measurements of many small molecules). The relatively poor scattering and the inward displacement of the hydrogen atom are apparent.

Neutrons, by contrast, interact insignificantly with the electron density when they pass through a crystalline solid. Scattering is instead by the nuclei. Since both nuclei and neutrons are extremely small, significant scattering takes place only when a neutron passes close to a nucleus, and on average the total intensity of diffraction of neutrons by a crystal is low compared with that of X-rays. The relatively weak scattering means that larger crystals are preferred for neutron diffraction, and it may not be easy to grow them. On the other hand, neutron scattering by a stationary atom does not fall off at higher angle like that of X-rays (Fig. 4.2); lower intensities at higher angles are due entirely to atomic vibrations. The weaker atom–neutron interactions also mean that absorption of neutrons by

The intensity of neutrons available at modern spallation sources has greatly improved this situation, enabling the use of considerably smaller crystals than previously.

single crystals is usually negligible, even when heavy atoms are present, in contrast to the situation with X-rays.

Although the scattering power of an atom for X-rays is directly proportional to its atomic number (the number of electrons in the neutral atom), there is no simple relationship between neutron scattering power and atomic number. Neutron scattering powers vary erratically across the periodic table, often with large differences between adjacent elements, and heavier elements do not dominate lighter ones as they do with X-rays; even different isotopes of the same element have different **neutron scattering factors**. A selection of relative scattering powers for X-rays and neutrons is given in Table 4.1.

It can be seen that some nuclei scatter in phase (positive scattering factors), while others scatter out of phase (negative scattering factors). Note that different isotopes of the same element may have quite different neutron scattering powers; this is particularly so for the isotopes of hydrogen, H and D. Elements (isotopes) with very small scattering powers, such as V, are effectively almost invisible to neutrons. Among common elements, H (D even more so) and N are particularly good neutron scatterers. There are several important consequences of this difference in the nature of X-ray and neutron scattering, which make neutron diffraction a useful tool in particular cases.

Compared with X-rays, neutrons are generally good at locating light atoms in the presence of much heavier atoms, though this depends very much on the particular elements involved. In particular, the precise location of first-row atoms such as C, N, O in structures containing several very heavy atoms such as W, Re, U is likely to be more successful with neutron diffraction, though an X-ray result is perfectly adequate in most cases (see, for example, case study 1) unless small differences in light-atom bond lengths are to be detected.

An extreme case is, of course, the location of hydrogen atoms, for which neutron diffraction is far superior to X-ray diffraction, especially for deuterated compounds. Not only is the neutron result more precise, because H/D atoms scatter

The use of neutrons for diffraction is somewhat more complicated than the brief treatment provided here. For example, scattering that is **inelastic** (change of wavelength on scattering) and/or incoherent (more complicated phase relationships) needs to be recognized and handled correctly.

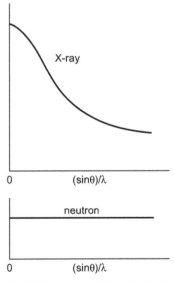

Table 4.1 Relative X-ray and neutron scattering factors of selected elements and isotopes. The two sets of values are not on the same scale; neutron scattering is much weaker.

Atom	X-ray	Neutron
H	1	−3.7
D	1	6.7
C	6	6.6
N	7	9.4
O	8	5.8
^{35}Cl	17	11.7
^{37}Cl	17	3.1
V	23	−0.4
W	74	4.8
Re	75	9.2
U	92	8.4

Fig. 4.2 The variation of X-ray (top) and neutron (bottom) atomic scattering factors with Bragg angle for stationary atoms. In both cases atomic vibration causes a reduction in scattering factors at higher Bragg angles.

The distinction between precision and accuracy is important (in all of science!) **Precision** refers to the spread of results obtained if a measurement is repeated many times; it measures repeatability or the degree of confidence with which a particular measurement can be made, and it is measured by statistical parameters such as s.u.s. **Accuracy** refers to the agreement of the measurement or result with the true (usually wanted but unknown) value. Thus, a result can be precise but not accurate (like a wrongly set digital clock), and it can be accurate but not precise.

Under certain circumstances, significant differences between the X-ray scattering factors of neighbouring elements can be generated by choosing a wavelength which gives a large resonant scattering effect for one of them and a small effect for the other. This requires tuneable X-ray wavelengths, which can be achieved with synchrotron radiation but not with standard laboratory sources.

relatively strongly, it is also more accurate, because it locates the nuclei directly rather than the electron density distorted by valence effects. For studies in which precise and accurate hydrogen atom location is important, neutron diffraction is the method of choice. Examples include hydride (H^-) ligands in transition metal complexes, bridging hydrogen atoms in electron-deficient compounds such as boranes, and unusual hydrogen bonding. In the majority of structures, however, hydrogen atom positions are entirely predictable and neutron diffraction is not justified.

Neutron diffraction can clearly distinguish many pairs of neighbouring elements in the periodic table, which have almost the same X-ray scattering power. This may be of value in some compounds such as mixed-metal complexes (e.g. containing both W and Re, which have 74 and 75 electrons, respectively, but quite different neutron scattering powers), alloys (where metal atoms may be ordered or disordered), and minerals.

Distinguishing between isotopes of the same element is impossible with X-ray diffraction but, in many cases, straightforward with neutrons, provided the isotopes are not disordered in the structure. A case in point is the determination of the H and D sites in a partially deuterated compound, which may help, for example, in establishing a reaction mechanism by unambiguously identifying the isotopic substitution in the product.

It should also be noted that neutrons have a magnetic moment, which interacts with the magnetic moments of atoms containing unpaired electrons. In paramagnetic materials, the atomic moments are randomly oriented, so these effects are averaged out and there is no extra information available with neutrons. Ferromagnetic, ferrimagnetic and antiferromagnetic materials, on the other hand, have an ordered arrangement of atomic moments, which often leads to an increase in the size of the unit cell when this effect is included; neutron diffraction produces extra diffraction maxima corresponding to the larger unit cell (a supercell), and can thus characterize the magnetic ordering in such compounds.

There are more advanced types of experiment, in which both X-rays and neutrons are used to study the same structure. Since neutrons locate nuclei, from which core electron density can be calculated, and X-rays reveal the total electron density distribution, the combination provides a means of mapping valence electrons and bonding effects. Such approaches (so-called 'charge density studies') require extremely careful measurements and corrections, since the valence effects are small compared with the total electron density, and they lie beyond the scope of this book.

4.3 Diffraction by powder samples

A single crystal gives a diffraction pattern (with either X-rays or neutrons) with discrete diffracted beams, each in a definite direction relative to the orientation of the crystal and the incident beam, according to the Bragg equation. Because the diffraction conditions are severe, a stationary single crystal gives very few reflections (see Section 2.3). In order to generate the complete diffraction pattern it is necessary to rotate the crystal in the X-ray or neutron beam.

If several single crystals of the same material in different orientations are irradiated simultaneously by X-rays, each of them gives its own diffraction pattern and these are superimposed. As the composite sample is rotated, any particular reflection will be generated by each of the individual crystals at different times as the Bragg equation is satisfied; the Bragg angle and intensity will be the same in each case (assuming equal sizes of crystals), but the direction of the diffracted beam will vary, while always being inclined at 2θ to the straight-through direction (Fig. 4.3). On a flat detector perpendicular to the incident beam and on the opposite side of the sample, this set of corresponding reflections from the multiple crystals appears as identical spots on a circle. With an increasing number of identical and randomly oriented crystals, more such spots appear, all lying on the same circle, which is where a cone of diffracted radiation hits the detector. A microcrystalline powder consists of an essentially infinite number of tiny crystals and this produces a complete circle for a particular reflection.

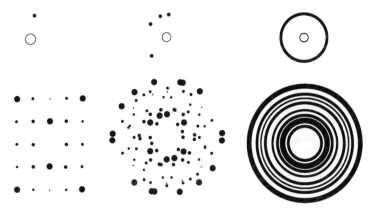

Fig. 4.3 The relationship between single-crystal and powder diffraction. **Top**: the effect for one individual reflection. The centre of the detector is marked with an open circle. Left: the position of one reflection from a single crystal. Centre: the positions of this reflection derived from four crystals together in different orientations. Right: the effect for a very large number of crystals. **Bottom**: the effect for a complete simple diffraction pattern. Left: the pattern from one carefully aligned single crystal. Centre: the patterns from four crystals superimposed in random relative orientations. Right: the pattern for a very large number of crystals; this is a powder diffraction pattern, and each spot in the left diagram has generated a complete circle in the right diagram.

The same occurs for every Bragg reflection, each one giving a cone of radiation with semi-angle 2θ (Fig. 4.4), and hence producing a circle on the detector. The overall result is a set of many concentric circles, with radii dictated by the Bragg equation and hence the unit cell geometry, and with intensities closely related to those that would be produced by one single crystal.

In practice, a powder diffraction pattern is usually measured either on a strip of photographic film wrapped round the sample in a cylindrical shape (a powder camera), in order to reach high Bragg angles (with θ approaching 90°, the

Even individual tiny crystals which can be seen only under a microscope, such as constitute a fine powder, are still effectively infinite in size compared with the wavelength of X-rays, so each one acts as a single crystal. In a fine powder, the number of individual crystals is also effectively infinite, with all possible orientations present simultaneously.

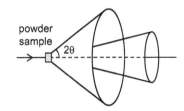

Fig. 4.4 Cones of diffracted X-rays produced by each reflection from a microcrystalline powder sample.

diffracted beam is almost doubled back on the incident beam), or by an electronic detector which is driven in a circle around the sample under computer control (a **powder diffractometer**). In either case, intensity is recorded as a function of angle, and for each reflection a Bragg angle and an intensity can be obtained. The effect of using a microcrystalline powder instead of a single crystal is to compress the full three-dimensional diffraction pattern into a one-dimensional pattern (the only geometrical variable is θ). It is also possible to use a position-sensitive detector (either an area detector, or one which is one-dimensional rather than two-dimensional in form) to record powder diffraction patterns quickly and efficiently.

For relatively simple structures, giving few reflections, there may be little overlap of these in the powder diffraction pattern. In such cases it is possible to assign individual indices and intensities and carry out structure determination as with single crystals. Even for larger structures and those of lower symmetry, where this is not possible, powder diffraction has important uses, in chemical analysis (both qualitative and quantitative) and for the identification of materials, which is its most common application. The determination of crystal structures from powder diffraction data, using advanced techniques for indexing the pattern to obtain a reliable unit cell, for solving the structure without a complete set of non-overlapping reflections (a problem with some similarities to twinning), and for refinement using the total measured powder diffraction profile instead of individual indexed reflection intensities, is a subject now undergoing rapid and impressive development. A proper treatment lies far beyond this text. The primary purpose of this brief treatment is to illustrate the relationship between single-crystal and powder diffraction techniques, both of which can be carried out with laboratory X-rays, synchrotron radiation, or neutrons.

4.4 Biological macromolecular crystallography

It was mentioned in Chapter 1 that a wide range of sizes of crystal structures can be determined by diffraction methods, from small molecules and simple salts to biological macromolecules such as proteins, nucleic acids, and assemblies including viruses and the ribosome. This book has focused on organic, inorganic, and organometallic compounds of interest to chemists; here we describe a few ways in which the technique is different when applied to large biological molecules. The principles are just the same, but they need to be used differently because of the nature of the samples themselves and the way in which they interact with X-rays.

In most cases there is a very limited amount of material available (and there may be significant safety requirements if it is a potent biological agent). Obtaining suitable single crystals can be a serious challenge, especially as the proteins are normally surrounded by a large amount of water within the crystals and they can be grown only from aqueous solution, often within a relatively narrow pH range, so there are fewer options for crystallization procedures, variation of the solvent itself not being one of them except to the extent that water-miscible

The earliest biological applications of X-ray crystallography were to proteins, and the term 'protein crystallography', abbreviated PX, was commonly used. With extension to nucleic acids and virus structures, the broad research area is now widely known as **macromolecular crystallography** or MX, and we will use this abbreviation here; for simplicity we shall also refer to 'proteins', with the understanding that this term includes also other macromolecules.

organic solvents such as alcohols can be added. Various additives (for example, polyethyleneglycols, inorganic salts, and ligands binding to the protein) can also be used to change the solubility, and other conditions that can be varied include the concentration and temperature. Special techniques have been developed to encourage crystallization from very small volumes of solution, and large arrays of nanolitre-scale crystallization cells can be set up in a combinatorial approach, possibly with robotic control of the varied conditions.

The significant separation of protein molecules from one another in the crystal means that it can generally be assumed that they have a molecular structure essentially the same as that in aqueous solution, and hence probably *in vivo*; if this were not the case, determining the crystal structure would be far less useful.

Crystals of biological macromolecules can be very beautiful in appearance, but they are often mechanically fragile and require special techniques for selection and mounting, along with some mother liquor, typically in small fibre loops or cavities in thin non-crystalline polymer holders looking rather like old-fashioned pen nibs; separation of crystals from the mother liquor before X-ray diffraction usually leads to serious degradation of crystal quality.

Unit cell axes are much longer than for most chemical compounds (generally in the hundreds of Å), and the larger cell volumes have two major impacts on X-ray diffraction patterns. First, there is a much greater density of reflections, the geometry of the diffraction pattern (the reciprocal lattice) being inversely related to the lattice geometry. Second, according to equation 2.1, the relative scattering power of a crystal is inversely proportional to the square of the unit cell volume, and approximately directly proportional to the sum of f^2 for all the atoms in one unit cell. While the $\sum f^2$ proportionality suggests strong diffraction from the large number of atoms, these are predominantly light atoms with small f values, and the inverse-square dependence on unit cell volume works strongly in the opposite direction, leading to diffraction that is weaker than that observed for chemical compounds with crystals of similar size. This general weakness is exacerbated by the extensive disorder that usually affects the solvent content of the crystal structure (and often some protein side-chains); as we have seen earlier, disorder reduces the intensities of reflections, especially those at higher Bragg angles. As a consequence, significant diffraction is often unobservable beyond a resolution (lattice plane d value) of 2 or 3 Å using laboratory X-ray sources, in contrast to well-behaved chemical systems for which a resolution considerably better than 1 Å is normal. This low resolution has impacts on data collection and reduction, and on structure solution and refinement, as well as limiting the detail with which the structural results can be described, because individual atoms, with bond lengths around 1.5 Å, are not clearly resolved from each other.

Techniques are currently being developed for growing crystals directly on a suitable support so that they do not need to be selected and mounted as a separate step.

MX samples tend to suffer greatly from **radiation damage**, the high energy of X-rays disrupting chemical bonds and generating radical species, especially ·OH radicals, which attack the protein molecules; this leads to decomposition and also to increased structural disorder. The effects are usually much reduced by collecting data at low temperature, but even then it is often necessary to use several crystals to obtain a full set of data, or to irradiate sequentially different parts of a single crystal with a very narrow (microfocus) X-ray beam, moving on to a new part as each becomes too damaged.

Very significant increases in lifetime of individual crystals are observed at temperatures down to around 100 K, which can readily be achieved by routine use of gas stream apparatus based on evaporated liquid nitrogen with a boiling point of 77 K; there is usually little or no further improvement at lower temperatures.

For the various reasons outlined above, synchrotron radiation is very widely used in MX: it provides high intensity even in a microfocus beam, and allows rapid data collection with modern pixel detectors. The availability of a wide range of X-ray wavelengths is also an advantage in providing data for 'anomalous dispersion' structure solution methods, as described later in this section. The experimental setup is quite similar to that used for chemical samples, but there is often only one rotation axis for the crystal (for simplicity), area detectors tend to be larger because they need to be positioned further away from the crystal to avoid overlap of the reflections in the dense diffraction pattern, and automation (including robotic sample mounting and removal) and remote control are common features to maximize throughput.

The data reduction step has to handle very large numbers of reflections, possibly measured from several crystals, so scaling of the different contributions is important. Unless the experiment is with a heavy-atom derivative as an aid to structure solution (see later in this section), absorption is not usually a significant effect. Because biological macromolecules are chiral and present as only one enantiomer (proteins contain naturally occurring amino acids and nucleic acids contain naturally occurring sugars), the only possible space groups are those which have only proper rotations (simple or screw); inversion, reflection, and other improper symmetry operations cannot occur. There are 65 of these, the so-called **Sohnke space groups**.

This is true also, of course, for a single enantiomer of any chemical compound; in synthetic chemistry, however, it is possible for partial or complete racemization to occur in subsequent stages, even after an enantiomerically pure compound has been obtained.

Standard methods of solving crystal structures of chemical compounds, described in Chapter 2, are rarely successful in MX. Most proteins do not contain heavy atoms, and if they are introduced in a derivative or into the solvent region of the structure, they do not usually dominate the scattering enough for routine Patterson map interpretation and subsequent Fourier-based structure completion. The probabilities of phase relationships in conventional direct methods depend on the size of the structure, becoming smaller as the size increases, so these methods are insufficiently reliable; they also do not work well if atomic resolution is not achieved in the data. Additional information is needed to solve the phase problem in MX, leading to the development of methods different from those already described. Three methods are particularly popular.

Isomorphous replacement depends on obtaining one or more additional crystal structures in which a heavy atom has been introduced in the asymmetric unit while leaving the rest of the structure essentially unchanged; the space group must be the same, and the unit cell parameters sufficiently similar for the structures to be considered isomorphous (within about 1%). The heavy atom may be, for example, a small mercury compound or a salt of the $[PtCl_4]^{2-}$ anion. The diffraction patterns of these isomorphous crystals will be similar but not identical, the scattering by the heavy atoms making an overall small but significant contribution. The difference between the two patterns is the heavy atom scattering and hence the heavy atom can be located in a Patterson map using the squares of the differences between the amplitudes measured for the native and derivative structures, or by variants of direct methods. Estimates of reflection phases can then be obtained by comparing the two sets of amplitudes and knowing the true heavy-atom contribution to each reflection, the amplitude and phase of which can be calculated by a Fourier transform from the known heavy atom position. With only one isomorphous derivative, ambiguities arise, but these can be resolved if there are at least two heavy-atom derivatives with the heavy atoms

The protein without incorporation of heavy atoms is called the native structure and has measured reflection amplitudes $|F_P|$; the amplitudes for an isomorphous heavy-atom derivative are $|F_{PH}|$. The squared differences $(|F_{PH}| - |F_P|)^2$ are used to calculate a 'difference Patterson map' from which the heavy atom position is found. The map may not be very clear, because the difference $\|F_{PH}| - |F_P\|$ is only an approximation for the heavy-atom diffraction amplitudes $|F_H|$; calculation of the true heavy atom contribution requires a vector (complex number) difference, $F_H = F_{PH} - F_P$, which involves phases as well as amplitudes.

in different positions (the method is then called multiple isomorphous replace-ment or MIR, in contrast to SIR).

The more generally applicable **molecular replacement** method does not require isomorphous derivatives, but it does depend on having a known crystal structure, possibly from a database, for a protein that is believed to have a similar molecular structure to that being investigated. This is taken as a model struc-ture for the unknown one, but its orientation and location in the unit cell need to be determined. Essentially, this is the same as the Patterson search method described in Section 2.7, though the detailed implementation may be different because of the large size of the structure.

The third method for solving MX structures uses resonant scattering, also known as anomalous dispersion, which leads to a breakdown of Friedel's law, with reflections h, k, l and $-h, -k, -l$ having different intensities (see Section 2.10). It is best performed with data from a synchrotron source, at which the wavelength can be tuned precisely to give the desired effects. A **multiple-wave-length anomalous dispersion** (MAD) experiment measures data at three different wavelengths, chosen such that one gives the maximum value of the imaginary scattering factor component f'' for a particular element in the structure, one gives the maximum value (whether positive or negative) of the component f', both of these lying close to what is called an absorption edge of the element, and the third is well away from the edge so that resonant scattering components are small. The same crystal (or crystals taken from the same batch if this is not pos-sible) is used for all three measurements. The three data sets serve essentially the same purpose as three isomorphous derivatives in the MIR approach, with the advantages that only one crystal is required, there is no doubt about the isomor-phous relationship, and the information content is higher because there are also Friedel pair differences; the resonant scatterer in the structure plays the same part as the heavy atom in the isomorphous derivative. Measurement of a single set of data at the wavelength of maximum f'' (single-wavelength anomalous dis-persion, SAD) is equivalent to SIR, with the Friedel differences as large as possi-ble. Heavy atoms are often used as resonant scatterers, but significant and useful effects can be obtained from atoms considerably lighter than those needed for MIR and SIR, for example first-row transition metals. Selenium also has a strong resonant scattering effect for commonly used wavelengths and can be intro-duced into proteins in multiple locations by substitution of selenomethionine for the sulfur-containing amino acid methionine. If suitable heavy atom deriva-tives are available, it is also possible to combine the techniques of isomorphous replacement and 'anomalous scattering' to provide even more information to help find reflection phases.

When MX diffraction data are available to less than atomic resolution (i.e. larger d_{min} than about 1.1–1.2 Å), electron density maps, even with completely correct phases, do not show individual resolved atoms, but rather continuous chains of electron density. The interpretation of such maps is, therefore, less straightfor-ward than for chemical samples, and involves computer graphics-aided manipu-lation of molecular models to fit the calculated electron density. This is less of a problem than it might be, because the individual structural components (amino

MAD may also be understood as an abbreviation for multiple-wavelength (or multiwavelength) anomalous diffraction, and likewise for SAD. Note that resonant scattering is principally used as a reflection phasing tool in MX, whereas its main use in chemical crystallography is for determination of absolute configuration; this is not necessary in MX, as naturally occurring amino acids and sugars have known enantiomeric forms.

Such techniques, not surprisingly, have their own acronyms: SIRAS and MIRAS for single and multiple isomorphous replacement with anomalous scattering, respectively.

acids, nucleobases, sugars, and phosphate groups) have well-known and largely invariant bond lengths and angles, and even some of the torsion angles have a tendency to fall within certain recognized ranges. Highly disordered solvent is often found in protein crystals (though hydrogen bonding does impose some order, especially in interactions with the protein molecules themselves), and this is treated in a similar way to the method described in Section 2.10.

The weakness of high-angle data means that MX structures generally have a much lower ratio of reflections to atoms than chemical crystal structures, and so it is not possible to refine freely as many parameters. The use of restraints is common, and in many cases isotropic rather than anisotropic displacement parameters are refined. Because of the size and complexity of the structures, different graphical forms of representation are often used; a common model shows a protein back-bone chain as a ribbon, with differently coloured segments for the various amino acids or for different structural features in the protein folding, and this clearly displays important folding motifs such as α-helices and β-sheets (Fig. 4.5). Ligands bound to proteins can be highlighted, for example as ball-and-spoke or space-filling models attached to the ribbons, and important solvent interactions added.

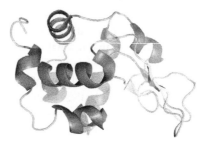

Fig. 4.5 A cartoon representation of one of many known crystal structures of lysozyme; only the protein molecule is shown, without any solvent. Compare this with the very different ball-and-spoke representation in Fig. 1.2.

MX structures have their own internationally recognized database, the PDB, already mentioned in Section 2.12; it is growing very rapidly. Structures are subject to validation in a similar way to those of chemical compounds, though the details are different.

4.5 Crystal structure prediction

Experimental chemistry, including spectroscopic and diffraction methods for structural characterization, are often complemented by theoretical calculations that seek to interpret and explain known results and predict as-yet unknown ones. Such approaches include molecular orbital calculations at a variety of levels of sophistication using *ab initio*, density functional, and other quantum theory methods, molecular modelling based on classical mechanics force fields, and molecular dynamics simulations. Many of these calculations are made for isolated single molecules, and refer essentially to behaviour at absolute zero, 0 K, though some methods provide for inclusion of a solvent environment.

Similarly, theoretical calculations can be made to predict or rationalize crystal structures. The task is considerably more challenging than for a single molecule, because intermolecular as well as intramolecular interactions have to be considered. The basic aim is to find crystal structures that have a lattice energy as low as possible. A simple approach would be to find the minimum-energy conformation of a single molecule by one of the methods mentioned above, then to try various arrangements of this in different space groups. Unfortunately this is far from sufficient, as intermolecular interactions can stabilize a less favourable molecular conformation and, in any case, the molecular structure will usually be perturbed significantly in a crystal environment, so the conformation itself must be allowed to change in the lattice energy minimization. Further complications arise from the possibility of including solvent molecules to generate solvate structures, the possible presence of disorder (not easily handled), the need to consider structures with $Z' > 1$ (considerably increasing the amount of calculation required), and the existence of potential polymorphs and temperature-dependent or pressure-dependent phase transitions, which may be indicated by finding two or more crystal structures with similar lattice energies. One way of assessing the relative merits of what are often many possible predicted crystal structures for a given compound is to produce a scattergram of lattice energy against calculated density; the most favourable structures are expected to have a minimum value for the first of these, and maximum for the second.

Crystal structure prediction is an expanding topic of current research interest, with important applications such as pharmaceutical polymorph screening, but it is very demanding of computing resources and it is difficult to ensure efficient and exhaustive coverage of all reasonable possibilities with so many variables.

4.6 **Summary**

- Single crystals can diffract neutrons with an appropriate wavelength given by the de Broglie relationship. The diffraction is much weaker than with X-rays, because the scattering is due to interactions with the very small atomic nuclei rather than with electrons. However, the neutron scattering from stationary atoms does not decrease with increasing Bragg angle.

- There is no simple pattern of neutron scattering factors for atoms across and down the periodic table, but an apparent random variation; most of the range of scattering factors is much smaller than the range of X-ray scattering factors, which are proportional to atomic number. Therefore most light atoms are more readily located in the presence of heavy atoms by neutron than by X-ray diffraction. This is particularly true for the location of hydrogen atoms, if such a result is of sufficient importance to outweigh the greater financial cost and technical difficulties associated with neutron diffraction.

- Neutrons are also generally more able than X-rays to distinguish between atoms of elements that have similar atomic numbers.

- Neutrons can distinguish different isotopes of the same element, which is not possible with X-rays.
- Neutrons have a magnetic moment, which interacts with ordered magnetic moments of atoms with unpaired electrons, providing additional structural information for such materials.
- Microcrystalline powders give diffraction patterns (with X-rays or neutrons) that are a superposition of large numbers of very weak single-crystal diffraction patterns. The three-dimensional information in a single-crystal diffraction pattern is thus compressed into one dimension, the only geometrical variable being the Bragg angle; a powder diffraction pattern is usually a profile trace of intensity against Bragg angle.
- Powder diffraction can be used as a 'fingerprint' analysis tool, providing both qualitative and quantitative (from relative intensities) information on single compounds and mixtures. In some cases it is possible to solve and refine crystal structures from powder diffraction data, and this technique is rapidly developing in power and reliability.
- Biological macromolecules diffract X-rays (and neutrons) in the same way as chemical compounds, but there are significant differences in the quality of diffraction patterns and in experimental procedures as a result of the much larger unit cells, high content of solvent that is usually disordered, and susceptibility of crystals to X-ray damage. Data are usually collected at low temperature with synchrotron radiation, and are usually obtained to a lower resolution than for smaller ordered structures.
- Special techniques are needed for solving and refining macromolecular structures, exploiting isomorphous relationships and resonant scattering effects in structure solution and extensive restraints in refinement. Large-scale biological structures are often displayed in quite different graphical styles from the familiar chemical molecular models.
- Crystal structure prediction (or rationalization for a known structure) involves extensive lattice energy calculations and variation of molecular conformations in response to intermolecular forces. Calculations are complicated by issues of polymorphism, solvate formation, and structures with chemically identical but crystallographically independent molecules in the asymmetric unit.

4.7 Exercises

Exercise 4.1

What advantages would there be in the use of neutron diffraction for crystal structure determination of each of the following, with one exception, and why would X-ray diffraction be preferable for that one case?

a) The product of a reaction of an organic compound with D_2O in a study of stereochemistry.

b) A polynuclear osmium carbonyl complex in which differences in the C–O
   bond lengths of terminal and bridging ligands is of importance.

c) A natural product containing C, H, N, and O for which the chemical identity
   needs to be confirmed.

d) An aluminosilicate mineral which may have the framework Al and Si atoms
   ordered or disordered.

e) A platinum complex of a boron hydride which may involve Pt–H–B bridging
   bonds.

Exercise 4.2

Why might neutrons be preferable to X-rays as the radiation source in the structural study of samples contained in special apparatus for controlling the sample
environment, such as some high-pressure cells or devices for maintaining a particular gaseous atmosphere? Why are air-sensitive crystals sometimes encased in
a vanadium capsule for neutron diffraction study?

Exercise 4.3

Why does a powder diffractometer usually include provision to rotate the sample around one axis, but only one?

Exercise 4.4

What difference would twinning make to a powder diffraction pattern?

Exercise 4.5

Use equation 2.1 to calculate the approximate relative scattering power of crystals of the following materials, assuming all the crystals to have the same volume;
this serves to illustrate some of the difficulty encountered in macromolecular
crystallography and the reason why even tiny crystals of simple compounds can
be studied readily with synchrotron radiation. For simplicity, consider all atoms
present to be carbon.

- Diamond, with 8 atoms in each cubic unit cell, $a = 3.57$Å.

- A benzene solvate of buckminsterfullerene, $C_{60} \cdot 4C_6H_6$ (ignore the H atoms),
  with $Z = 2$ in a triclinic unit cell of volume 2294 Å^3.

- A protein with about 300 amino acids (2750 non-H atoms), $Z = 4$, in an
  orthorhombic unit cell, $a = 50.1$, $b = 67.2$, $c = 92.2$Å.

Glossary

18 Å³ rule. A rough rule of thumb, applicable for most organic, organometallic, and coordination compounds, based on the observation that the average volume for non-hydrogen atoms in a crystal structure is usually fairly close to 18 Å³. This can be used to estimate the number of atoms (and hence molecules) in the unit cell at an early stage of a crystal structure determination.

Absolute configuration. The assignment of the correct enantiomer of a chiral molecule (or non-molecular chiral solid material).

Absolute structure. A general term encompassing absolute configuration for chiral materials and related properties (such as polarity) of achiral but non-centrosymmetric crystal structures; the correct choice between a structure and its inversion-related counterpart.

Accuracy. The agreement of a measurement or derived result with the true (usually desired but unknown) value. See also **Precision**.

Amplitude. The size of a wave, measured from zero deviation (the mean value, between maxima and minima) to the wave maximum. The amplitude of a beam of X-rays is proportional to the square root of the intensity.

Angstrom unit. More correctly, Ångström unit; a measure of distance equal to $100\,pm = 0.1\,nm = 10^{-10}$ m.

Anisotropic. An anisotropic property, or function, is one that has different values in different directions or orientations.

Anisotropic displacement parameter. A set of (usually 6) parameters describing the mean-square amplitude of an atom in a crystal structure in different directions.

Anomalous scattering (or anomalous dispersion). An alternative (and inappropriate) name for resonant scattering.

Area detector. A device for recording part or all of a diffraction pattern, such that the position of each reflection on the face of the detector is known as well as the intensity; in some cases, the time at which the reflection is recorded is also known.

Asymmetric unit. The unique, symmetry-independent portion (a rational fraction of the unit cell volume) of a crystal structure. Application of the space group symmetry operations to the asymmetric unit generates the complete crystal structure, and it is the asymmetric unit that must be determined using X-ray diffraction.

Atomic scattering factor (for X-rays). The variation in X-ray scattering power of an individual atom as a function of Bragg angle (usually expressed as a function of $(\sin\theta)/\lambda$).

Ball-and-stick (or ball-and-spoke). A commonly used model for molecules in which atoms are represented as spheres, connected by rods representing bonds.

Bond angle. The angle enclosed between two bonds B–A and B–C formed by a particular atom B.

Bond length. The distance (usually measured in Å, nm, or pm) between two atoms that are considered to be directly bonded to each other.

Bragg equation (or Bragg's Law). A single equation, with associated geometrical definitions and conditions, that describes X-ray diffraction by a single crystal, relating angles of diffraction to the indices and spacings of sets of parallel lattice planes: $\lambda = 2d_{hkl}\sin\theta$

Centred unit cell. A unit cell with lattice points at its eight corners and also at some or all of its six faces or at the centre of the unit cell; there is also a centred trigonal unit cell for rhombohedral crystal structures, which has lattice points at the positions $(\frac{2}{3}, \frac{1}{3}, \frac{1}{3})$ and $(\frac{1}{3}, \frac{2}{3}, \frac{2}{3})$. A centred unit cell is chosen so that it has the characteristic shape for its crystal system.

Charge flipping. A dual-space method of solving crystal structures in which the direct-space modification is the reversal of sign of all calculated electron density below a particular threshold value.

CheckCIF. An online tool provided by the International Union of Crystallography for the validation of a CIF.

Chiral. A molecule or other object is chiral if it is not identical to any conformation of its mirror image.

Co-crystal. A crystal structure containing two or more distinct chemical species. Note that the term co-crystal is not usually applied to solvates of a single main species.

Complex number. A quantity having two numerical components that may be considered as orthogonal to each other in some way and cannot be combined by simple scalar addition. Complex numbers are used in crystallography to represent reflections (as an alternative to explicit amplitude and phase) and in Fourier transform equations.

Conformation. Different conformations are different geometries of a molecule that can be interconverted by moving atoms relative to each other without breaking and making any bonds, for example by rotating about bonds in a chain or flexing a ring.

Constraints. Mathematical relationships applied rigidly to combinations of refined parameters (or imposed on individual parameters), such that the total number of independent refined parameters is reduced. Constraints are thus imposed on the model structure, possibly in conflict with the requirements of the diffraction data.

Crystal. A solid material that gives essentially a sharp diffraction pattern with most of the intensity in Bragg reflections. (This is the definition adopted by the International Union of Crystallography.)

Crystallographic Information File (also Format or Framework), CIF. An internationally agreed standard for the archive and exchange of crystal structure results (and other crystallographic information, including data and publication details) according to a CIF dictionary maintained and developed by the International Union of Crystallography.

Crystallography. A broad range of scientific theories and methods concerned with the study of solid materials, mostly but not exclusively crystalline in form.

Crystal structure determination. The application of diffraction methods to find the positions of atoms in the structure of a crystalline material.

Crystal systems. Seven different arrangements of types of symmetry element in crystalline solids, leading to seven characteristic shapes of unit cells. The seven systems are: triclinic, monoclinic, orthorhombic, tetragonal, trigonal, hexagonal, and cubic.

Database. A computer-based collection of items of information with a common structure and format, usually with associated software for its efficient management, searching, and manipulation of the contents. International crystallographic databases are comprehensive major research tools and repositories of published and deposited structural information.

Data reduction. The process of converting raw measured intensities to structure factor amplitudes (or their squares) by applying corrections for absorption, geometrical effects, and other factors concerned in the experiment.

de Broglie relationship. The relationship between the momentum p of a moving particle such as a neutron or electron and its associated wavelength λ with applications in properties such as diffraction: $\lambda = h/p$.

Difference electron density map. A reverse Fourier transform in which $|F_o|$ is replaced by $|F_o| - |F_c|$, so that atoms in the trial structure are suppressed in the resulting electron density map, and new atoms are revealed more clearly.

Diffraction. Cooperative scattering, involving interference effects, of radiation (electromagnetic or particulate) by a collection of objects such as molecules in a crystal structure.

Diffractometer. A device for rotating a single crystal to different orientations in an X-ray beam and recording the diffraction pattern on a detector, under computer control.

Direct methods. A general term encompassing methods for solving crystal structures from the measured diffraction amplitudes, using no other information except the known properties of electron density distributions and typical molecular geometry, which impose restrictions on relationships among phases of reflections with related indices.

Dual-space methods. Methods for solving crystal structures involving repeated forward and reverse Fourier transforms with successive modification of the direct-space information available (such as recognized features of molecular geometry and selection of candidate atoms) and reciprocal-space information (replacement of calculated by observed amplitudes, and possible application of probabilistic phase relationships).

Dynamic disorder. A term sometimes used to refer to atomic displacements, which increase with temperature, especially where these are unusually large.

***E* values** (normalized structure amplitudes). Structure factor amplitudes set on a normalized common scale by dividing each one by the average value for reflections with a similar Bragg angle, the data set being divided into ranges of Bragg angle to obtain an average for each range. E values are an estimate of diffraction amplitudes that would be measured for point atoms at rest, i.e. no spread of electron density for the atoms. (There are some symmetry-related factors also involved in this calculation.)

Electromagnetic radiation. A form of energy consisting of coupled oscillating electric and magnetic fields, covering a wide range from radio waves (long wavelength, low frequency) to X-rays and γ-rays (short wavelength, high frequency) and with a common and constant velocity in a vacuum.

Enantiomers. The two mirror images of a chiral molecule.

Ewald sphere. A geometrical construction used to demonstrate X-ray diffraction, which predicts the direction of each reflection and the orientation of the crystal relative to the incident X-ray beam at which it occurs.

Fourier transform. A mathematical relationship between two functions that have mutually inverse (reciprocal) dimensions. In crystallography, a diffraction pattern is the Fourier transform of a crystal structure, and vice versa.

Frequency (of a wave). The number of waves occurring in one second (units are $s^{-1} = Hz$).

Friedel's law. The equality of intensities of reflections with indices h, k, l and $-h, -k, -l$. This applies always for centrosymmetric structures, but only in the absence of significant resonant scattering effects for non-centrosymmetric structures.

General position. Any position in the unit cell of a crystal structure which does not lie on a pure rotation axis, mirror plane, inversion centre, or central point of an improper rotation axis. An atom lying in a general position is not transformed into itself by any of the space-group operations (except the trivial identity operation).

Glide plane. A symmetry element combining reflection with a translation component, a rational fraction of a relevant lattice repeat, in a direction within the plane.

Goniometer head. A device on which a single crystal is mounted for measurement of a diffraction pattern, providing lateral (and possibly angular) adjustments to enable the centring of the crystal in the X-ray beam.

Goodness of fit. A statistical parameter, closely related to R factors and incorporating standard uncertainties, that gives an overall indication of how well the calculated and observed amplitudes match. An ideal agreement and correct weighting scheme based on the s.u.s should give a goodness of fit equal to 1, and this should be the case for subsets of the data displayed over ranges of intensity, Bragg angle, and other variables.

Hydrate. A solvate in which the solvent is water.

Hydrogen bonding (and hydrogen bond). A significant attractive interaction between a hydrogen atom usually bonded to an electronegative atom, and another electronegative atom in the same or another molecule. Hydrogen bonds generally contribute attractive energies that are a small fraction of covalent or ionic bonds.

Indices. Three integers h, k, l, that specify a particular Bragg reflection (also known as reflection indices) and also a set of parallel lattice planes (also known as Miller indices, especially in describing external crystal faces that are parallel to these planes).

Inelastic scattering. Scattering (mainly of neutrons) in which the wavelength is changed because some of the radiation energy is lost to, or gained from, the scattering atom, which recoils.

Insertion devices. Complex arrays of many magnets inserted in the straight sections between bending magnets of a synchrotron storage ring, designed to generate X-rays with very high intensity and other special properties of interest.

Intermolecular. Between two (or more) molecules (referring to forces, interactions, etc.).

International Tables for Crystallography. A series of reference books, available in print and online, produced by the International Union of Crystallography and covering many aspects of the theory and practice of crystallography. *Volume A* is the standard reference for space-group symmetry.

Intramolecular. Within a single molecule (referring to forces, interactions, etc.).

Isomorphous and **isostructural.** Two (or more) crystal structures are isomorphous if they have very similar unit cell parameters and the same space group. If, in addition, the atoms lie in essentially the same positions in the two structures, they are isostructural.

Isomorphous replacement. A method for solving macromolecular crystal structures, in which data are collected for two or more isomorphous samples and the differences in the diffraction patterns are used to locate the heavy atoms incorporated in the isomorphous derivative(s) of the native substance.

Isotropic. An isotropic property, or function, is one that has the same value in all directions or orientations.

Isotropic displacement parameter. A single parameter describing the mean-square amplitude of an atom in a crystal structure.

Lattice. A set of points (lattice points) regularly spaced in one, two, or three dimensions, equivalent to each other by pure translation symmetry; a crystal lattice shows the repeating nature of the crystal structure but not the actual contents of the repeat structural unit.

Lattice parameters (or unit cell parameters). The three axis lengths (a, b, c) and three interaxial angles (α, β, γ) describing the geometry of a crystal unit cell.

Lattice planes. Sets of parallel planes passing through lattice points, with applications to crystal faces (morphology) and diffraction.

Laue conditions (or Laue equations). Three equations, one for each dimension, describing the diffraction of X-rays by a single crystal.

Macromolecular crystallography (MX). The term generally applied currently to the determination of crystal structures of biological macromolecules and their larger assemblies.

Microscope. An optical device with a combination of lenses for obtaining a magnified image of a small object. The use of electron diffraction with electromagnetic focusing of the scattered electrons to obtain a magnified image is known, by analogy, as electron microscopy.

Molecular replacement. The name given to the Patterson search method for solving crystal structures when it is used for biological macromolecules.

Monochromatic. Having a single wavelength (literally, 'single coloured', used of radiation).

Monochromator. A single crystal or other material that exploits the Bragg equation to select a single wavelength from an X-ray beam.

Mosaic spread. The small angular range of misalignment of sub-microscopic domains in a real, as opposed to perfect, single crystal.

Multiple-wavelength (and single-wavelength) anomalous dispersion. Methods for solving macromolecular crystal structures in which the resonant scattering of some elements is used to help locate those atoms; it involves measuring intensity differences for Friedel pairs of reflections, preferably at several different X-ray wavelengths with the same crystal.

Neutron scattering factor. The scattering power of a particular atomic nucleus for neutrons. Unlike X-ray scattering factors, they are independent of Bragg angle for stationary atoms, and they are different for different isotopes of the same element. They are much smaller (though usually measured in different units) than X-ray scattering factors.

Neutron spallation source. A pulsed source of neutrons (and other elementary particles) generated by impact of high-energy protons or other particles from a synchrotron with a heavy-metal target.

Oligomer. An association of two or more copies of a chemical species (which may or may not exist itself as a stable entity) in a single discrete molecule. Specific terms are used for different numbers of associated monomers: dimer, trimer, tetramer, etc. polymer.

Particulate radiation. The wave nature associated, through the de Broglie relationship, with moving elementary particles such as neutrons or electrons.

Patterson search. A method of solving a crystal structure by matching the set of interatomic vectors expected for a known (or assumed) molecular fragment structure to the Patterson function, while rotating and then translating the fragment to its correct orientation and position in the asymmetric unit.

Patterson synthesis (or Patterson map). A reverse Fourier transform in which $|F_o|$ is replaced by $|F_o|^2$ and all phases are set to zero. The result is a map of vectors between all pairs of atoms in the crystal structure, from which the positions of some (usually heavy) atoms can be found in some cases, providing an initial solution of the phase problem.

Phase (of a wave). The position of the maximum of a waveform (measured along the direction of the wave) relative to some defined origin; usually only relative phases are important (when waves interact), so the choice of origin is unimportant. The phase may be measured as a dimensionless fraction of the wavelength, or as an angle such that one whole wavelength corresponds to $360°$ (2π radians).

Phase problem. The loss of relative phase information for individual reflections when a diffraction pattern is recorded on a detector, thus providing only the direction and intensity (related to amplitude) of each reflection. Direct reconstruction of an image of the structure by Fourier synthesis is impossible without the missing phases.

Point group. The complete set of symmetry operations for a molecule or other single finite object.

Polymorphs. Two or more different crystal structures of the same chemical compound, or the same solvate. Note that different solvates, even solvates containing different amounts of the same solvent, are not strictly polymorphs; they are sometimes (confusingly) called pseudo-polymorphs.

Powder diffraction. The diffraction of X-rays or neutrons by a microcrystalline powder sample, in which the three-dimensional diffraction pattern, including orientational information, is compressed into a one-dimensional pattern with the Bragg angle as the only geometrical variable by superposition of huge numbers of identical weak single-crystal patterns.

Powder diffractometer. A device for recording a powder diffraction pattern, with the main components of a radiation source and monochromator, powder sample, detector (single element, one-dimensional position-sensitive, or area detector), and facilities to rotate the sample and detector.

Precision. A measure of the spread of results obtained if a measurement is repeated many times, or an estimate of this spread obtained from statistical analysis of a single measurement. Precision in crystallography is expressed by standard uncertainties. See also **Accuracy**.

Primitive unit cell. A unit cell with lattice points only at its eight corners.

Quasicrystal. A crystalline material giving a sharp diffraction pattern as a result of short-range structural order but lacking long-range periodic order in the form of a lattice.

Radiation damage. The deterioration in crystal quality and diffraction pattern observed for some samples during X-ray irradiation, particularly for biological samples. Radiation damage tends to be much lower with neutrons.

Reciprocal lattice. A lattice with dimensions of $Å^{-1}$ uniquely related to the crystal (direct) lattice, used to describe the geometry of a diffraction pattern.

REFCODE. A unique identifier of 6–8 characters assigned to each entry in the Cambridge Structural Database (CSD).

Refinement. The process of systematically adjusting the numerical parameters of a trial structure (mainly atom positions and displacement parameters) to make the calculated diffraction amplitudes match the observed amplitudes as closely as possible, as assessed by a least-squares discrepancy definition.

Refraction. The alteration in the direction of travel of light as it passes from one medium into another with a different refractive index.

Refractive index. The ratio between the velocity of light (or other electromagnetic radiation) in a vacuum and its velocity in a particular medium.

Resolution. The minimum lattice plane d-spacing in a measured diffraction pattern, corresponding to the maximum Bragg angle. This corresponds to the smallest interatomic distance that can effectively be resolved in the crystal structure. True atomic resolution requires a d-spacing resolution of about 1.1 Å or better (lower).

Resonant scattering. A modification to normal X-ray scattering by an atom that occurs when the X-ray photon energy is close to a value appropriate for promotion of an electron from one orbital of the atom to another, or for complete removal (ionization) of an electron. The amount of resonant scattering depends on the element and on the X-ray wavelength.

Restraints. Mathematical relationships among refined parameters, representing reasonable expected behaviour for molecular geometry, atomic displacements, etc., that are formulated as additional 'experimental observations' alongside the diffraction data in structure refinement. Restraints, in contrast to constraints, do not reduce the number of refined parameters, and are used in coordination with, rather than in potential opposition to, the diffraction data.

***R* factors** (or residual factors). Various numerical functions providing single-value assessments of how well a trial structure accounts for the observed diffraction pattern. The sum of differences between observed and calculated amplitudes (or some function of these, possibly weighted to reflect perceived reliabilities) is divided by the sum of observed amplitudes and expressed as a simple number or a percentage. *R* factors decrease as a trial structure is improved during structure solution and refinement. There are also *R* factors, defined in an analogous way, for assessing the internal agreement of a set of diffraction data containing symmetry-equivalent reflections.

Rhombohedral. A subset of the trigonal crystal system, in which the primitive unit cell may be regarded as a cube that is either compressed or elongated along one of its four body diagonals; this diagonal retains its threefold rotation symmetry as the trigonal axis. A conventional trigonal unit cell has three times the volume of the primitive rhombohedral cell, with two additional lattice points (see **Centred unit cell**).

Screw axis. A symmetry element combining rotation with a translation component, a rational fraction of the relevant lattice repeat, along the axis direction.

Single crystal. A crystal in which all the unit cells are identical and aligned in essentially the same orientation, thereby generating a clear single diffraction pattern.

Sohnke space groups. The 65 space groups that have no improper symmetry elements—no inversion-rotation axes, reflection, or inversion. Materials consisting of only one enantiomer, including biological macromolecules, can crystallize only in these space groups.

Solvate. A crystal structure containing solvent of crystallization in addition to the main component(s).

Solvent of crystallization. Molecules of solvent, used in the synthesis or crystallization of materials, that are incorporated in the crystal structure.

Space-filling. A model of a molecule in which atoms are represented by intersecting spheres with the appropriate van der

Waals (non-bonding) radii, and bonds between the atoms are not visible.

Space group. The complete set of symmetry operations for a crystal structure. There are 230 space groups.

Special position. Any position in the unit cell of a crystal structure which lies on a pure rotation axis, mirror plane, inversion centre, or central point of an improper rotation axis. An atom lying in a special position is transformed into itself by at least one of the space group operations other than the identity operation.

Spectroscopy. The probing of energy levels of a material by measuring the absorption or emission of radiation in order to investigate aspects of structure or carry out qualitative or quantitative analysis.

Standard uncertainty (s.u.). A statistically derived estimate of the precision of a measurement or calculated result, also known previously as estimated standard deviation as it represents an approximation to the expected spread of measurements or results if the experiment were to be repeated many times.

Static disorder. A random variation in the detailed contents of the asymmetric unit, involving alternative positions for some atoms and/or a mixture of different atom types on a common site. This appears in the trial structure as partially occupied atom sites in an average asymmetric unit.

Structure factor. A term used for the combination of amplitude and phase for a particular reflection in a diffraction pattern.

Supramolecular. A term literally meaning 'above or beyond molecular', usually applied to assemblies of separate molecules having specific interactions with each other that can be predicted or rationalized.

Synchrotron. Charged elementary particles (usually electrons or positrons) confined by magnets to an almost circular (actually polygonal) path and travelling at relativistic velocities emit a broad range of electromagnetic radiation, known as synchrotron radiation, at each magnetically induced change of direction (bending magnets and insertion devices). The output includes X-rays of very high intensity and with special properties different from those of laboratory-generated X-rays.

Systematic absences. Subsets of reflections in a complete diffraction pattern, or in particular sections or rows of it, that have zero intensity as a result of translation symmetry elements in a space group (unit cell centring, glide planes, or screw axes).

Tautomeric forms (tautomers). Different forms (isomers) of a molecule related to each other purely by the concerted migration of electrons (in bonds and lone pairs) and hydrogen atoms.

Torsion angle. The angle between two bonds (B–A and C–D) formed by two directly bonded atoms (B and C) when viewed in projection along the B–C bond joining these two atoms; the angle is positive if B–A must be rotated clockwise around B–C to make its projection coincident with that of C–D.

Translation. A form of symmetry consisting of effectively infinite repetition of a basic unit in the same orientation with regular spacing in one, two, or (for crystalline materials) three dimensions.

Trial structure. A model structure containing some, or all, of the atoms of a crystal structure in what are currently believed to be approximately their correct positions. Solving and refining a structure consists of gradually completing and improving the trial structure until its calculated diffraction pattern matches the observed one as closely as possible.

Twin fraction. The relative amounts of two (or more) components in a twinned crystal.

Twin law. A 3×3 matrix defining the relative orientations (with or without inversion) of two components of a twinned crystal structure.

Twinning. The presence of two or more orientations or mirror images of the same crystal structure in a well-defined geometrical relationship to each other in a crystal sample, leading to a superposition of the individual single-crystal diffraction patterns.

Unique set (of data). The total set of reflections, up to a particular maximum Bragg angle, that are independent of each other by symmetry. The unique set of data in reciprocal space (the diffraction pattern) corresponds to the asymmetric unit in direct space (the crystal structure).

Unit cell. The basic structural unit of a crystal structure; repetition of the unit cell at each lattice point generates the complete crystal structure.

Unit cell parameters. See **Lattice parameters**.

Vector. A quantity having both magnitude and direction. Vectors are used extensively in crystallography, including to represent atom positions (as distance and direction from the unit cell origin) and reflections (with amplitude and phase). A quantity having magnitude but no direction (such as volume) is called a scalar.

Wavelength. The separation, in distance units (usually Å in crystallography), between adjacent maxima of a wave.

Weights. In the calculation of Fourier maps and least-squares refinement, individual reflections may be assigned weights according to their perceived reliability. Weights are usually based on $1/\sigma^2(F_o^2)$ or a similar function, with possible incorporation of additional terms depending on intensity, Bragg angle, and other variables.

X-ray absorption. The absorption of X-rays as they pass through a crystalline sample, whether they are simultaneously diffracted or not. Absorption reduces the intensity of diffracted X-rays and is dependent on path length, so it is different for different (even symmetry-equivalent) reflections, and a correction must be made if it is significant.

X-ray camera. A device (now largely superseded) for recording diffraction patterns on photographic film.

X-ray tube. A laboratory source of X-rays involving the high-energy impact of an electron beam with a metal target, leading to the ejection of core electrons and relaxation of electrons from higher-energy atomic orbitals; the orbital energy differences are in the X-ray range of the electromagnetic spectrum and are characteristic of the target metal.

Z and Z′. Z is the number of chemical formula units (molecules, etc.) in one unit cell of a crystal structure. Z' is the number of chemical formula units in the asymmetric unit.

Further reading

The following books provide more detailed accounts or different approaches to the subject suitable for the target readership of this Primer. This is not intended to be an exhaustive list; some classic crystallographic texts are not included, because they are inappropriate at the undergraduate chemistry level or because they are rather outdated in content or approach, and the list is deliberately kept short.

Blake, A.J., Clegg, W., Cole, J.M., Evans, J.S.O., Main, P., Parsons, S. and Watkin, D.J. 2009. *Crystal Structure Analysis: Principles and Practice*, 2nd ed. Oxford: OUP.

Blow, D. 2002. *Outline of Crystallography for Biologists*. Oxford: OUP.

Glusker, J.P. and Trueblood, K.N., 2010. *Crystal Structure Analysis: A Primer*, 3rd ed. Oxford: OUP.

Hammond, C. 2015. *The Basics of Crystallography and Diffraction*, 4th ed. Oxford: OUP.

Massa, W. (English translation by Gould, R.), 2004. *Crystal Structure Determination*, 2nd ed. Berlin: Springer.

Index